Science Beyond the Classroom Boundari
for 7–11 Year Olds

Science Beyond the Classroom Boundaries for 7–11 Year Olds

Lynne Bianchi and Rosemary Feasey

Illustrations by Simone Hesketh

Open University Press

Open University Press
McGraw-Hill Education
McGraw-Hill House
Shoppenhangers Road
Maidenhead
Berkshire
England
SL6 2QL

email: enquiries@openup.co.uk
world wide web: www.openup.co.uk

and Two Penn Plaza, New York, NY 10121–2289, USA

First published 2011

A catalogue record of this book is available from the British Library

ISBN-13: 978-0-33-524132-3 (pb) 978-0-33-524133-0 (hb)
ISBN-10: 0-33-524132-8 (pb) 0-33-524133-6 (hb)
eISBN: 978-0-33-524134-7

Library of Congress Cataloging-in-Publication Data
CIP data applied for

Typeset by RefineCatch Limited, Bungay, Suffolk
Printed in the UK by CPI Antony Rowe, Chippenham and Eastbourne

The McGraw-Hill Companies

To my daughters, Eve and Kate, for their lust for life, questioning attitudes and their genuine smiles and laughter, love Mummy.

To my parents-in-law, Eve and Frank, for being so wonderful and such enthusiasts for the outdoors. Rosemary

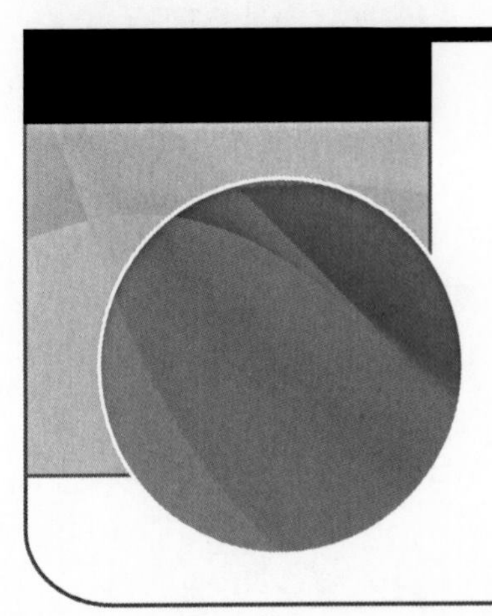

Contents

List of figures

List of tables

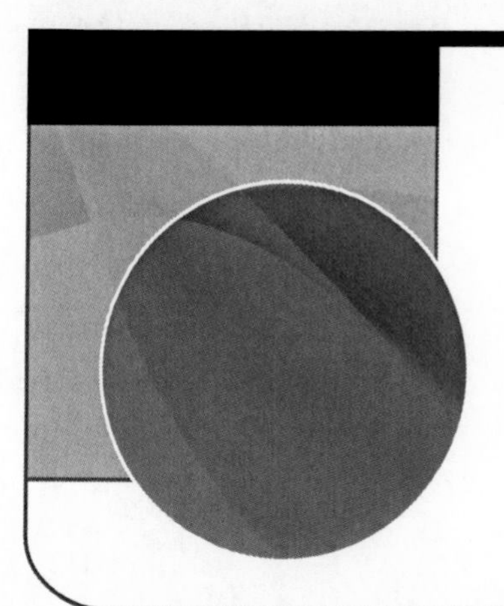

List of illustrations

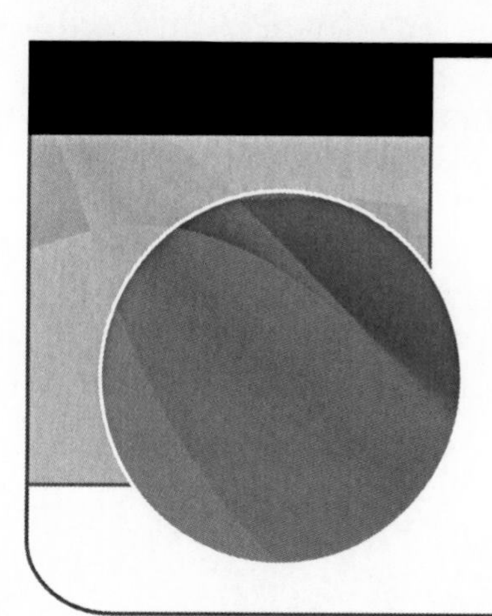

Acknowledgements

We are indebted to the following schools and teachers who have been so generous with their time and in sharing their wonderfully creative practice with us for the purpose of this book. We would also like to thank the many teachers we have met along the way who have been kind enough to allow ideas to be shared. You are all an inspiration.

Bradway Primary School, Sheffield

Bradway Primary School (formerly known as Sir Harold Jackson Primary School) is a large primary school located in the south-west of Sheffield surrounded by beautiful countryside less than a mile away from the border with Derbyshire.

Castleside Primary School, Consett, Co. Durham

This small primary school is tucked away in a housing estate in Castleside, in a small village situated two miles south-west of Consett.

Diamond Hall Junior School, Sunderland

This is a large junior school that is situated in a socially and economically deprived area close to Sunderland city centre where the children and staff use the outdoors, especially their enclosed courtyard.

Fox Hill Primary School, Sheffield

This is an inner-city school with children of mixed ability included an integrated resource, they would describe themselves as welcoming, a friendly school and a fun place to be.

Gainford CE Primary School, Co. Durham

This is a small primary school with around 90 children in a charming village on the north bank of the River Tees. The school was built in 1857 so space is limited but the school grounds offer lots of potential for work outdoors.

Grenoside Community Primary School

This is a school with a 330-pupil intake to a new school with grounds ripe for development. The school is on the north side of Sheffield surrounded by a varied landscape of woodland, parks, village and town life.

Monteney Primary School

Monteney is an inner-city school in Sheffield; a lively and fun place for both staff and children to learn and achieve. Pupils enjoy a varied and creative curriculum.

Shaw CE Primary School, Shaw, Wiltshire

This is a small semi-rural school with a mix of new and old buildings and the where the school grounds are being creatively developed for science outdoors.

St. Thomas More

This is a small catholic primary school in the north of Sheffield, set in beautiful grounds and ideal for outdoor learning.

Wheatlands Primary School, Redcar, Cleveland

This school is on the southern outskirts of the seaside town of Redcar and has limited room inside but lots of room in the school grounds for doing science.

Woolley Wood Special School, Sheffield

This is a primary school for children with severe learning difficulties and profound and multiple learning difficulties. The school is situated in the north of Sheffield and has a wide catchment area, taking children from 2½ to 12 years old.

Our heartfelt and sincere thanks go to Lynne's dad, who so kindly read through the draft manuscript and corrected our silly sentences and offered suggestions for changes. (Love you Dad – Lx)

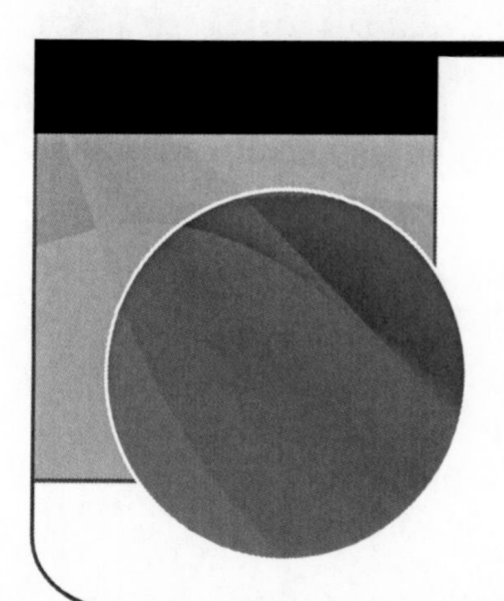

Introduction

Science 'Beyond the Classroom Boundaries'

This is a truly innovative book that aims to support schools to change the parameters of teaching and learning in primary science, by changing teacher perceptions of where science should be taught. Underpinning this is the resolve to develop children's Personal Capabilities (Bianchi, 2002) so that eventually children are able to choose why and when to take their learning 'Beyond the Classroom Boundaries' on a regular basis. This is not a traditional approach in primary science but a bold step which results in a whole-school approach to moving the science curriculum outdoors into the school grounds.

What do we mean by 'Beyond the Classroom Boundaries'? In the context of this book we will focus on the school grounds (although you might wish to consider the idea that 'Beyond the Classroom Boundaries' might include, corridors, school hall, library, ICT room, etc.) as the main environment for teaching and learning in primary science. It is not our intention to go beyond the school grounds, since there are many organizations such as 'Forest Schools' and Outdoor Education Centres supporting work in the wider environment, but little has been developed that illustrates how almost the whole of the science curriculum can be developed using the immediate outdoor environment.

The aim is to build on the good practice begun in the early years, and to ensure that children's Personal Capabilities are further developed in relation to their understanding and skills in science by ensuring that children regularly work 'Beyond the Classroom Boundaries' rather than almost exclusively in the classroom.

Robert Brown MSP, Deputy Minister for Education and Young People, offered a thought-provoking question when he commented: 'We must challenge people to think: Why learn indoors?' (Available online at www.eriding.net/educ_visits/learning.shtml accessed 25 June 2010).

'Beyond the Classroom Boundaries' offers an innovative view of how, when and where children learn in science. We have been fortunate to be able to draw upon the experiences and approaches developed by practising teachers from across England, who have been generous in sharing their ideas and expertise, so that the reader can appreciate how the rhetoric of working 'Beyond the Classroom Boundaries' in science has been translated into reality across the primary years.

Each chapter in this book is structured into four sections, enabling the reader to consider what the aims to be achieved (the goals), what the current reality in schools is (the reality), suggested options for development (the options) and possible ideas for what you will be able to do in your school (the will). This structure draws on the established 'GROW' model which is a technique for problem solving or goal setting, developed in the UK by Graham Alexander, Alan Fine and Sir John Whitmore and is often used for the purpose of coaching. The value of GROW is that it provides an effective, structured methodology which both helps set goals effectively and is a problem-solving process.

Stages of GROW

There are a number of different versions of the GROW model. This version presents one view of the stages but there are others.

Table I.1 The GROW model

G Goal	This is the endpoint, where you want to be. The Goal has to be defined in such a way that it is very clear to the child when you have achieved it.
R Reality	This is how far you are away from your goal. If you were to look at all the steps they need to take in order to achieve the goal, the Reality would be the number of those steps they have completed so far.
O Options	Sometimes also including an exploration of the obstacles, this stage asks you to identify ways of dealing with them and making progress. These are the Options.
W Will or Way Forward	The Options then need to be converted into action steps which will take you to your goal. These are your 'Wills' or Ways Forward.

References

Bianchi, L. (2002) Teachers' experiences of the teaching of Personal Capabilities through the science curriculum, Ph.D. Thesis, Sheffield University.

For more information on the GROW model visit www.en.wikipedia.org/wiki/GROW_model.

CHAPTER 1

Transforming science 'Beyond the Classroom Boundaries'

What is the goal?

There have been many years of innovation in primary science education. Surprisingly though, most of this has taken place within the confines of the classroom. What primary science has not yet done with universal success is to step outside the classroom boundaries to use the school grounds for teaching and learning across the science curriculum.

> There is a long history of what is taught there, by whom and how, as well as embedded perceptions about the types of buildings and spaces where these practices traditionally occur. However, there is a need to challenge existing assumptions . . . this requires taking a different approach to imagine a whole range of different possibilities.
>
> (Futurelab, 2008: 12)

At this point we should make the distinction that the focus of 'Beyond the Classroom Boundaries' is the 'school grounds' and not the more traditional routes of learning beyond the classroom, such as, through educational visits to museums, walks around the community and exploring local settings. This book is different in focusing on developing most areas of primary science outside the classroom but within the boundaries of the school grounds.

The aim of this chapter is to begin to challenge the perceived notions of where children should experience ways of thinking and working 'Beyond the Classroom Boundaries', what it means to learn as a primary scientist, and indeed what it means to learn as a whole by embracing the development of personal skills and capabilities. Our view is that we can transform science across the primary years by exploring these three avenues.

Alexander (2009) suggests a curriculum which includes aims such as:

- well-being
- empowerment
- autonomy
- independence
- fostering skills
- exciting the imagination
- empowering local citizenship.

You will find that in this book many of the above aims form the basis of our pedagogical rationale that is predicated on the concept of science across the school environment, such that children develop as scientists through constant and consistent engagement with both the indoor and the outdoor school environment.

Primary science needs to offer children appropriate contexts for understanding about how they learn and about the science itself. The classroom is not ideal for this; in fact, we would suggest that the school classroom offers only a sanitized version of science and one in which many children have limited opportunities to develop independence and to use and apply their science in a range of real-life contexts.

Cars down a ramp

OK, most of us have taught this activity, where children explore friction in the classroom, by putting different surfaces on ramps and sending cars down to find out how different surfaces affect how far the car travels. The usual materials used for the surfaces are things like bubble wrap, sandpaper, carpet, linoleum, etc.

The following question helps to make the point of how inappropriate this activity really is. . .

- When was the last time you drove your car on bubble wrap, sand paper, carpet, linoleum . . .?
- Why would we want to contrive surfaces, when, within most school grounds we can give them access to a range of appropriate surfaces, from grass to tarmac, sand to soil, paving stones to rubberized surfaces?

Using the school grounds for science provides children with an environment that is closer to real-life experiences and everyday applications of science than in the classroom.

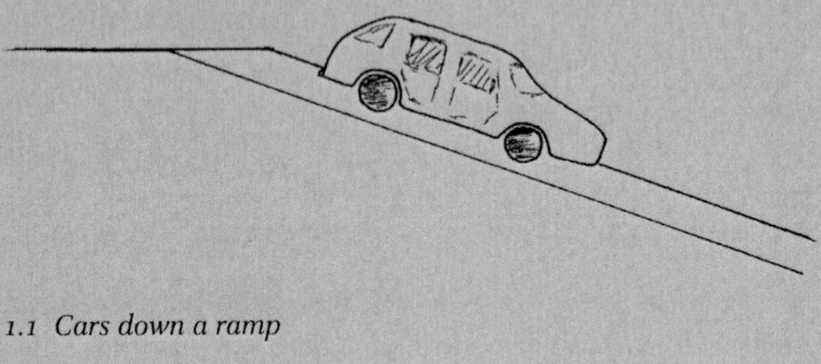

1.1 Cars down a ramp

The whole school, both indoors and outdoors, should be the theatre for learning in science and learning about oneself. This acknowledges that learning should focus on what is readily available in the school environment, to teach the components of physics, biology and chemistry, from cars down a ramp, to food chains and materials to micro-organisms.

Throughout this book an equally important element is the development of children's Personal Capabilities (Bianchi, 2002): namely, problem solving, creativity, communication, teamwork and self-management. As a co-author of this book, in her work Lynne has explored how the proactive use of these personal skills and capabilities can help develop children as increasingly scientifically capable and responsible learners, who have:

- an improved understanding of 'how' to learn
- better learning of science
- greater motivation to learn science
- increased engagement in learning science
- improved personal responsibility when learning science.

This complements idea of 'Learnacy' as promoted by Claxton (2004: 37):

> We should be helping them to develop supple and nimble minds, so that they will be able to learn whatever they need to. If we can achieve that, we will have a world-class work-force comprising people who are innovative and resourceful. The personal argument converges on the same conclusion. Many young people are patently floundering in the face of all the complexities and uncertainties of contemporary life.

So, what is it that we hope to achieve with this book? Well, first, that eventually all schools will provide a primary science curriculum where the children share responsibility for deciding whether to learn indoors or outdoors. Second, to develop understanding about how to help children work outdoors successfully to allow science learning to flourish by supporting them to work autonomously with the development of their personal capabilities. Our goal in this book is to help you in realizing these ambitions for your school.

What is the reality?

The reality is that in most primary schools, children in the early years (3–5 years old, Foundation Stage) are most likely to spend between 50 and 90 per cent of their time outdoors, usually moving between the indoors classroom and the external environment at their own choice.

In the early years a well-planned outdoors environment and the expectation that children will go outside in most weathers is the norm, not the exception. The children are in charge of their own movement and learning, and the role of the adult is to provide a watchful eye for safety purposes, to offer, where appropriate, quality intervention to scaffold learning and make a range of assessments. As children move through the primary years this changes dramatically, until in the final year of their primary career, their time outdoors shrinks to around 10 per cent and in some schools to nil. In other year groups there are limited opportunities to work outdoors, usually in contexts that are heavily supervised and scripted, and few of the activities outdoors are science events.

Interestingly, when pupils were asked about science in the outdoors, their responses included:

'We don't do much science outside, we do it mostly indoors.'

Alfie (age 8)

'We should because all the years we have been here we have covered most indoor activities, and outside there is lots more to learn.'

Toby (age 10)

Where schools have ventured outdoors beyond the early years classes, activity is frequently based around the school garden, or vegetable plot, where the focus is on tending and growing vegetables and some flowers. Often this activity is carried out by a small group, usually the gardening or Eco-club and not as an integral part of science lessons, where children explore and investigate plants, plant growth, habitats or issues relating to sustainability. Most science outdoors is based upon concept areas related to living things, with little or no focus on other areas such as the physical or material world, and even less on thinking and working scientifically.

The school grounds have immense potential for all kinds of learning, as Learning Though Landscapes acknowledges:

> School grounds are essential to children's learning and development, providing opportunities for healthy exercise, creative play, making friends, learning through doing and getting in touch with the natural world. We believe all children have the right to enjoy and benefit from well designed, managed and used school grounds.
>
> (www.ltl.org.uk/ accessed 26 August 2009)

What are the options?

Our options are quite clear: move primary science 'Beyond the Classroom Boundaries' and use the school grounds so that children develop and apply their Personal Capabilities when thinking and working in science. The alternative is to maintain the status quo, which, in our opinion, would be to restrict the opportunities and to limit the potential for learners, denying them a creative curriculum and good-quality primary science. Of course, we are clear in the option we would choose!

In learning science 'Beyond the Classroom Boundaries' not only will children use and develop Personal Capabilities in science, but they will also develop an appreciation that learning can take place in many different contexts.

Underpinning this is the hope that as children mature they develop a sympathetic awareness of their immediate environment and a commitment to considerate use of the school grounds. Along with this we would want children to take on personal responsibility for the school grounds as part of their daily school life, a sense of responsibility that we would hope will be extended to their lives outside the school.

What will you do?

In reading and using this book, we hope to support the reader to:

- develop creative and innovative practices and resources to support learning 'Beyond the Classroom Boundaries' that assist the development of children's Personal Capabilities in an embedded way
- transform the idea of learning spaces available for primary science
- explore the idea of mobile learning – that is, pupils can learn inside and outside the classroom on a 'needs must' basis
- consider the voice of the learner in involving children in the decision of what, when, how and where to learn in science
- develop systems for safe working practices for children working 'Beyond the Classroom Boundaries'
- understand how scientists and their approaches to science can be adapted and applied to working in the school environment
- be involved in changing the learning relationship where adults and children are engaged in more collaborative decisions in science and have autonomy to choose ways of working and learning.

None of this will happen overnight, and it will require teachers to audit science 'Beyond the Classroom Boundaries' and then prepare a plan of action to manage changes over an appropriate timescale. Subsequent chapters in this book aim to explore the issues related to science 'Beyond the Classroom Boundaries' and to offer support to teachers, taking into account the following suggestion from Learning Through Landscapes.

> Making physical changes to enable this to happen doesn't have to be expensive, and relatively small improvements outdoors can have disproportionately significant benefits ... The key to successful improvements outdoors is to focus on what you want the children to be able to **do**, rather than what you want them to **have**. Children's needs are diverse at this age, so a flexible and adaptable outdoor environment is the key to supporting their development.
>
> (www.ltl.org.uk/ accessed 26 August 2009)

The final section of this chapter, and indeed most chapters, offers practical ideas for science 'Beyond the Classroom Boundaries' which we hope will encourage the reader to move from the rhetoric of outdoor learning in science to reality.

Practical ideas for teaching forces 'Beyond the Classroom Boundaries'

So often children are engaged in activities to develop understanding about forces within the confines of the classroom, when everything that we need to teach about forces is, in fact, outside in the school grounds and free. This means less organization of resources, a wider range of materials and objects for children to work with and to develop ideas about forces in real-life situations. So let us have a look at how we could teach forces differently. The following section offers suggestions for developing key concepts related to forces and Personal Capabilities.

Forces and water

When beginning to work 'Beyond the Classroom Boundaries', it is sometimes easier to take the first step by allowing the children to work directly outside the classroom door. Alternatively, you could organize the class so that small groups of three or four children take a bowl or aquarium and a bag of equipment to explore forces and water outside. Before you go out ask the children to create a code of conduct for working which will guide them and help reinforce independence and responsibility for learning. Ask children to include how they will work together, communicate, manage themselves and think while outdoors: in essence, how their Personal Capabilities would best be demonstrated.

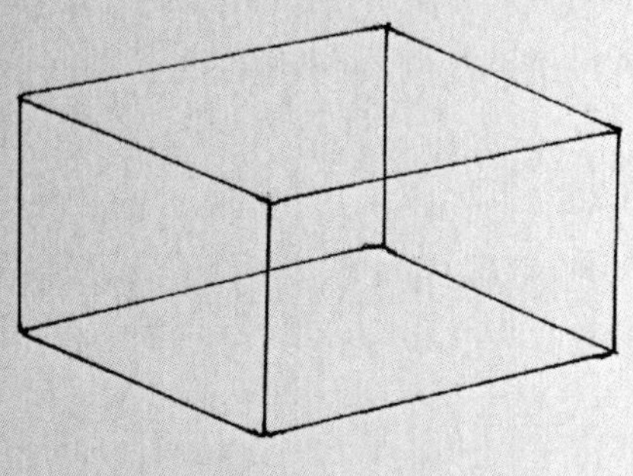

1.2 Aquarium

Balloons filled with air, water, ice

Suggest that the children try to push the balloon filled with air into the water. They will feel the water pushing back. Tell children to push the balloon into the water then let go; they will observe that the water pushes the balloon upwards.

Ice balloons will float because the ice is less dense than water. Sketch ice cubes and ice balloon.

What do you want to find out about ice in water?

1.3 Filled balloons

Cans of cola

What are the similarities and differences in the way they float? Why?

The ordinary coke will float slightly lower in the water than the diet cola because of the high sugar content which makes it more dense than the diet cola.

1.4 Cola cans

1. What will happen to the grapefruit when you put it in water: first with the peel then without?
2. What do you think will happen?
3. Why do you think it will happen?

A whole grapefruit with the peel will float because the peel is full of tiny holes that are full of air. Hence, it is less dense than the water; therefore it will float. Take the peel off and the grapefruit now is more dense, and will not float as high.

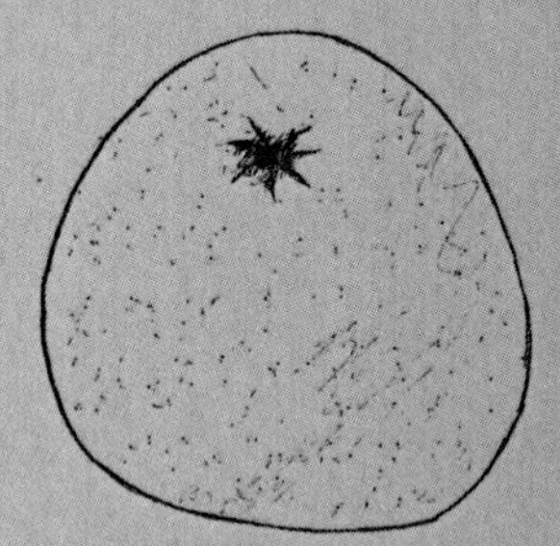

1.5 Grapefruit

Force of water and moving air – Whacky Boat Race

There are obvious reasons why this activity should take place 'Beyond the Classroom Boundaries', not only because water is being used but also because it takes up space in the classroom and it is not easy to dismantle if tables are required for another lesson. Setting up the guttering outside provides the context for:

- effective questioning
- planning and carrying out fair test investigations
- designing and making boats
- working collaboratively
- delegating roles
- making decisions of how to record observations and data.

Start the activity by giving children time to explore one of the boats and how to make it move down the gutter waterway. Then challenge children to ask a range of questions using different question stems. In relation to working 'Beyond the Classroom Boundaries' different question stems could be:

- placed on an outdoor white board
- painted onto a wall
- hung on trees, fences and so on.

The question stems should include:

how	what	what if
why	which	how does
when	would	where will

1.6 Sail shapes

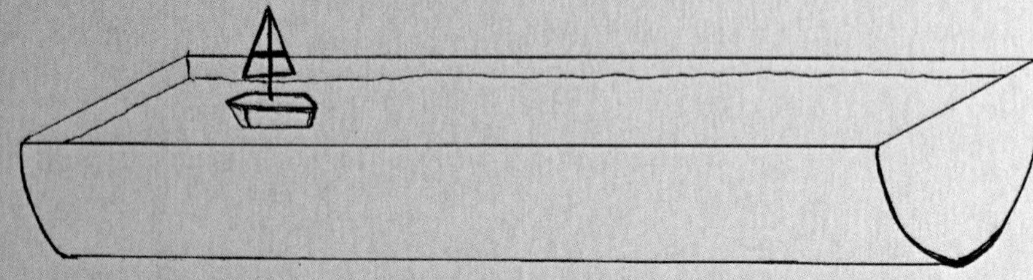

1.7 Guttering with boat

Of course, when children work outside we do have to make sure that they are working at an appropriate level. When children are given the opportunity to work independently the teacher is free to work with children carrying out formative assessments and, where appropriate, to challenge children's ideas and ways of working. Table 1.1 suggests a progression in children investigating sailing boats when children work in small groups or teams.

1.8 Sail shapes

Table 1.1 Progression in sailing boats

Progression in scientific enquiry	Progression in teamwork
Simple comparison by eye • Which boat goes the furthest?	Take turns, share and co-operate in small groups. Volunteer to take on a job to help others in a team. Show fairness to others. Help others and work at being a good friend, at times putting them first before yourself.

Progression in scientific enquiry	Progression in teamwork
Non-standard measures, simple table • Which boat carries the biggest load? • Which shape sail is the best? • Which material makes the best sail? **Standard measurement table, graph** • Which shape boat travels the furthest? • Which shape sail makes the boat travel the furthest? • Which sail material makes boat travel the furthest? • Which boat carries the heaviest load?	
Repeat readings • How does the amount of load affect how far the boat travels? • How does the position of the mast and sail affect how far it travels? • How does the position of a load affect the boat? • How does the area of the sail affect how far it travels? **Repeat readings, calculate speed, using subject knowledge to explain lone graph and conclusions** • How does the area of the sail affect the speed of the boat?	Co-operate and collaborate in small and larger groups. Take on given roles and responsibilities to help the team achieve a goal. Talk about how what you do in the team influences others. Be fair and compassionate, recognizing people's feelings and ideas. Identify what the team has done well and possible improvements. Begin to realize that people express feelings in different ways.

Forces – the water cascade

A water cascade is usually found in the early years setting and is rarely used in the later primary years. However, it is a fabulous piece of equipment for older children using either water or, as in this example, a dry cascade for sending cars or balls down, so that they explore different aspects of forces, such as:

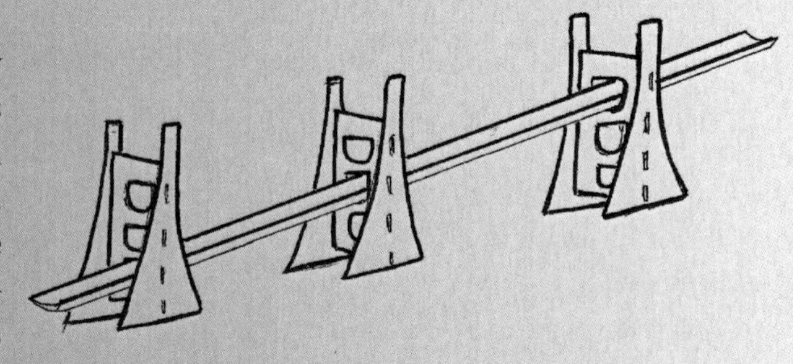

1.9 Water cascade

- a push or pull force can make an object speed up (move)
- friction can make an object slow down or stop
- we can change the direction in which something is moving
- gravity is a force that pulls an object towards the Earth.

When introducing a new piece of equipment such as the water cascade children of any age will need time to explore how to use it and the different ways it can be configured before they work on activities such as fair test investigations. When allowed time to explore children will extend their own experiences and also their science skills, subject knowledge and personal capabilities.

Table 1.2 Progression in using the water cascade

Progression in using the cascade	Progression in Personal Capabilities – communication
What can we put in the channel that will travel down the cascade? How can we get an object to go down the water cascade in less than 30 seconds?	Share opinions, feelings and ideas with others in a sensitive manner. Talk about issues of concern and give suggestions for action. Actively listen and respond to others by asking questions, checking for understanding. Adapt to different group members and when needing to present ideas to a wider audience.

Progression in using the cascade	Progression in Personal Capabilities – communication
How quickly do different objects travel down the water cascade? What is the reason for the differences in times? How can we show what we have been doing and explain the science behind it to other people?	Plan what to say so that ideas and lines of thinking can be understood by others. Actively listen and respond to others, asking questions, giving feedback or suggestions. Use negotiation as an effective forms of communication.

References

Alexander, R. (ed.) (2009) *Children, their World, their Education: Final Report and Recommendations of the Cambridge Primary Review*. Oxford: Routledge.

Bianchi, L. (2002) Teachers' experiences of the teaching of Personal Capabilities through the science curriculum, Ph.D. thesis, Sheffield Hallam University.

Claxton, G. (2004) A discussion paper for the Qualifications and Curriculum Authority, November.

Futurelab (2008) *Reimagining Outdoor Learning Spaces: Primary Capital, Co-design and Educational Transformation*. Bristol: Futurelab.

QCA: Futures Meeting the challenge – a curriculum for the future: subjects consider the challenge. http://webarchive.nationalarchives.gov.uk

CHAPTER

2 Identifying the potential for science 'Beyond the Classroom Boundaries'

What is the goal?

In this chapter we attempt to show how easy it is to:

- identify how the grounds are currently used for science
- identify the potential of the school grounds for science
- involve children in developing the school grounds to support science
- develop an action plan to achieve the redesign.

A key issue here is the involvement of the whole school, including the children, since there is 'evidence to suggest that adults gain a far better understanding of children's capabilities and interests and are often surprised by the skills, aptitudes and resourcefulness of children involved in co-design projects' (Futurelab, 2008: 21). Children do not have the conceptual barriers of time, funding or feasibility that can often inhibit the imagination of teachers. Thus, since changes are directed at children using the school grounds, it would seem obvious that children should be involved in the process.

What's the reality?

One of the first steps is to identify how the grounds are currently used for science. Teachers who have worked with this project began by mapping out their school grounds and annotating their maps using Post-it notes or coloured pens; they identified areas that teachers already use for science, indicating who, when, why and how the areas were used.

This provided an important starting point and usually indicated:

- that even the most developed school grounds were underused
- use was seasonal: favouring the warmer months
- the school grounds were used more frequently by younger year groups
- use was linked to natural science, with few activities linked to the material and physical sciences

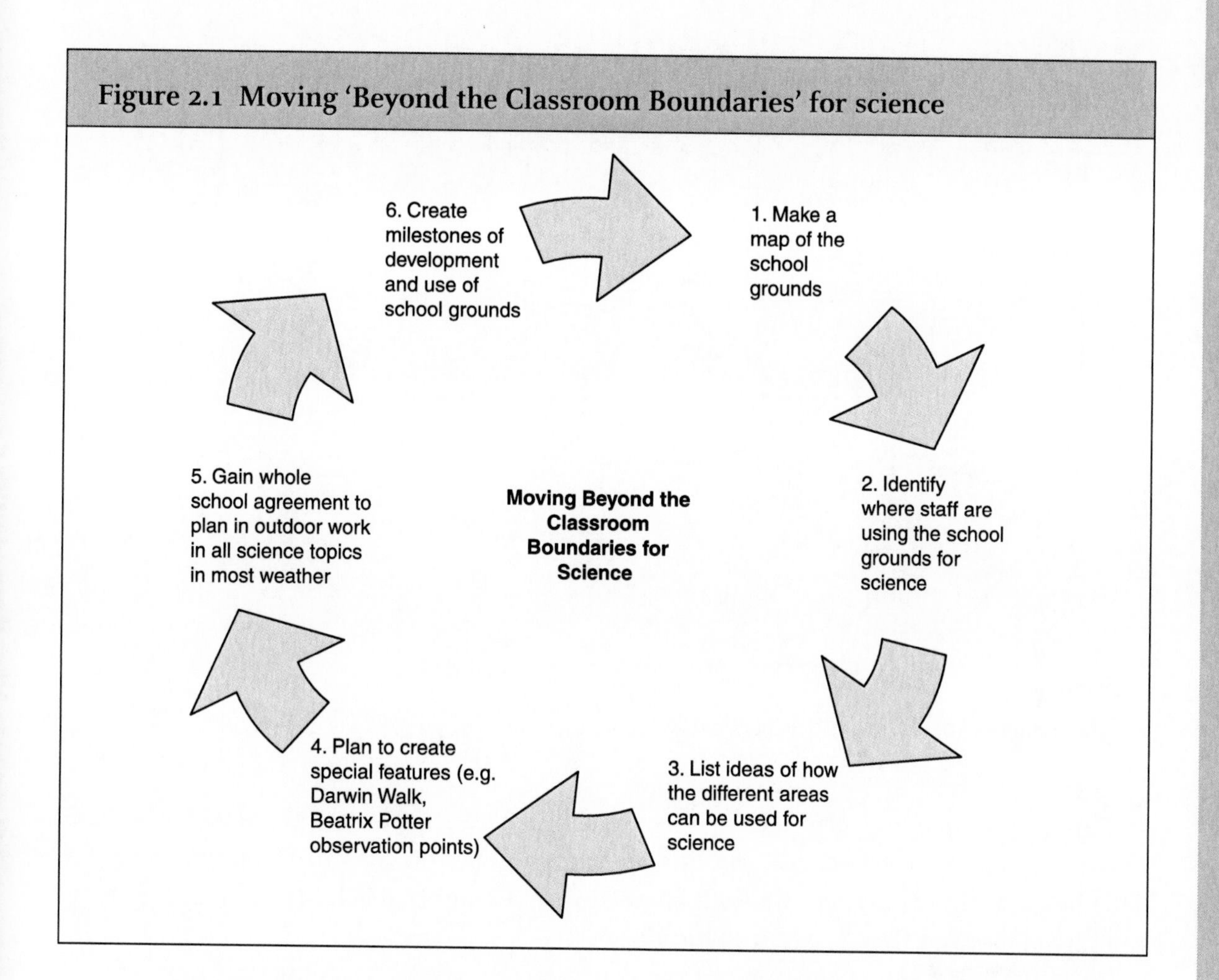

- activities were more observational than explorative or investigative
- that longitudinal studies were limited
- that activities were usually teacher-initiated and -controlled.

Auditing for science

At the beginning of developing science 'Beyond the Classroom Boundaries' Carol and her staff at Shaw Primary created a large map showing the current use of the school grounds for science. Gradually, they added ideas from everyone, so that pupils and visitors could see how their ideas were developing; these included initiatives such as the Darwin Thinking Walk, Science Zone and sitooteries.

From this point teachers were challenged to look again at the school grounds – this time in relation to teaching the majority of the science curriculum outside.

The aim of this was to support the staff in realizing that, in the first instance, most school grounds have potential that does not involve any expense and therefore is in fact free!

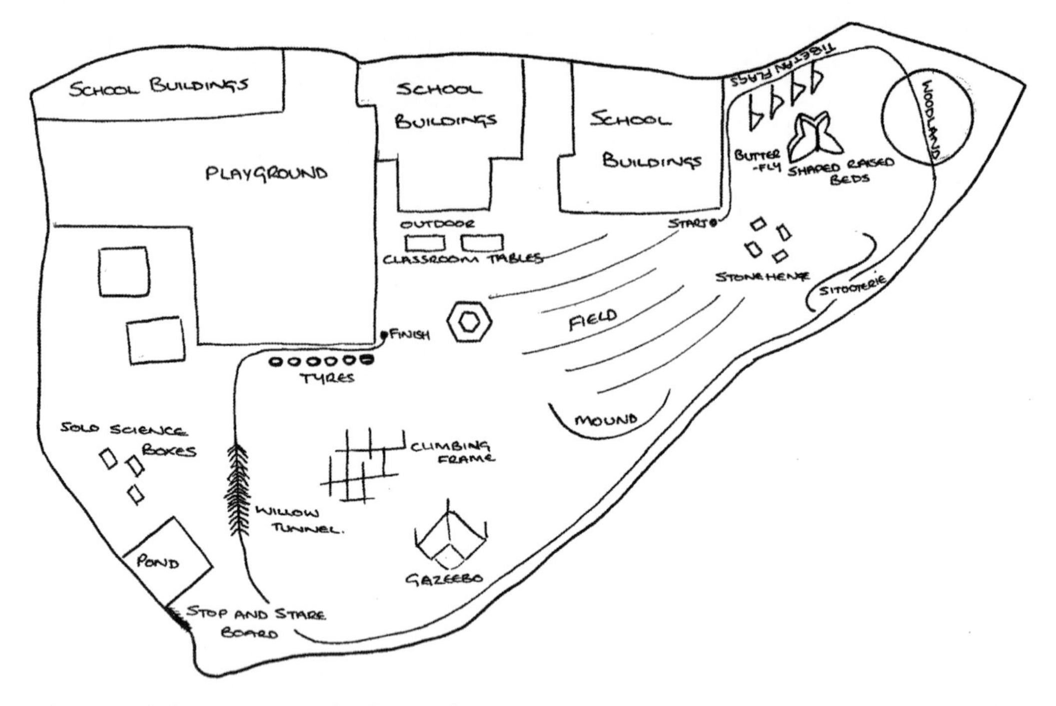

2.1 Audit map of Shaw Primary School Grounds

This is a simple, but effective activity where teachers are asked to consider what is already available in the school grounds and how to use the current resources in topics. This helps staff to realize how easy it can be to move science teaching and learning from indoors to 'Beyond the Classroom Boundaries'.

Once the staff appreciate the potential use of the school grounds, the next step is to consider how this can become a whole-school initiative and include the pupils in planning and decision making.

Auditing for Personal Capabilities

As part of this auditing process a group of teachers from Sheffield schools considered the ways in which children worked outdoors, as a whole class with teacher instruction and guidance, as a whole class observing demonstrations, as a whole class undertaking the same activity in smaller groups and in smaller groups with a 'find and bring back' or 'find and draw' approach. They considered their role while outdoors – crowd manager, instructor, timekeeper, facilitator, questioner and participant.

They considered how these methods of children and teacher working encouraged or discouraged the development of Personal Capabilities: teamwork, communication, creativity and problem solving and self-management. They SWOTed their current approaches (Strengths, Weaknesses, Opportunities and Threats) with regards to ways of working outside that could support the development of the personal skills and in turn promote autonomy, responsibility and greater independence when learning outdoors.

What are the options?

There are many ways that moving 'Beyond the Classroom Boundaries' can become a whole-school initiative and include the pupils in planning and decision making. For example:

- teachers working within or across year groups on how they use the school grounds for science
- class projects where children redesign the school grounds or a specific area for learning science outside
- the school council canvassing ideas from children, parents and staff, then presenting their ideas to the school
- inviting a science consultant to support the process.

Involving everyone

Schools such as Shaw, Grenoside and Castleside engaged staff and children in reconsidering the school grounds in relation to science. In doing so the children shared ideas and so were engaged in listening to each other, considering alternative ideas, debating advantages and limitations.

By including all staff (including teaching assistants and school governors) as well as children, their participation helped to develop a feeling of ownership and a sense of community.

What it also did was to encourage teachers in particular to rethink and possibly raise their aspirations in relation to what they and the children can do in the school grounds in science.

What would we like in the school grounds to support science?

In the north-east of England people have a saying 'Shy bairns get nowt!' and a very useful saying it is. If we take a conservative approach to the school grounds we will never help children to realize the potential for primary science. So the most important thing to do at this point is momentarily to put aside all issues relating to the outdoors, such as health and safety, money, staffing, challenging behaviour, and let your imagination run riot to create, on paper, your ideal school grounds for science.

Teachers and children who tried this found it very liberating and it led to a range of suggestions, such as those in Table 2.1.

Table 2.1 Teacher suggestions for the ideal outdoor environment

We would like	Making it real
Outdoor resource kits	Outdoor boxes containing, for example, binoculars, pooters, paper and pencils.
Wall murals	Painted on walls and could include identification keys for birds, invertebrates and trees. Solar system, water cycle, animal and plant life cycles.
Work stations	Using existing items such as logs, low walls and picnic type benches.
Beatrix Potter – 'sitooteries'	Logs, large rocks, seats designed and made by the children and local community.
Darwin 'Thinking Walk'	Designated walk around the perimeter of the school grounds with arrows, words and animal sculptures placed at points to denote the walk.
Work area	Table outside every classroom, perspex roof outside classrooms for shaded area for children to work.
Role-play areas outside	Shed outside each classroom which houses a role-play setting with tents outside and tables with role-play features such as geologists' station.
Pond	From new pond dug into ground to small tub pond using a barrel or other container.
Seating	Wooden seating designed by the children and built to be used across the curriculum for discussion, plays and demonstrations in collaboration with local community, for example, the Rotary Club.
Boards for writing	White or chalk boards for children to write or draw on, use as a notice board, observation board.
Hide for bird watching	A tent, donated by parents, camouflaged hide from science resource catalogues, purpose-built shelter (by friends of the school) with seating, board for logging observations and identification posters.
Bird tables/feeding stations/bird boxes	Designed and made by children and placed around school grounds.
Texture boards	Fixed to railings or walls with different materials for children to feel; alternatively, create texture 'posts' which identify different materials and textures around the grounds (e.g. metal, wood, brick, concrete).

We would like	Making it real
Mechanical systems	Mechanical objects and systems which show internal workings.
Bird box and webcam	Sited in school grounds and linked computer in position where children can access it inside the school.
Raised beds	Plant pots, tyres, unusual containers, wooden raised beds and railway sleepers.
Hanging baskets	Donated by parents.

If we engage teachers, pupils and the local community in developing the science potential of the school grounds, then we will be fostering engagement, ownership and responsibility where children are more likely to:

- use the facilities
- be interested in their environment
- apply science ideas and skills
- care for their environment
- use it to its full potential
- develop personal self-esteem
- develop pride in the school grounds.

Who can support us?

Schools involved with the project soon realized that once they had created their 'We would like list' the next step was to make their aims real, but of course where they could they wanted to achieve this in the most efficient and economical way. It would be naive to suggest that moving a school towards using the outdoors for science does not require some funding, but teachers recognized that they could request support from a range of people and agencies in the local community to achieve their goals in a cost-effective way.

Parents

Sarah at Castleside arranged for parents and friends of the school to donate days on a weekend where they tidied up the school grounds, removed bushes that were not needed and created new areas for learning including vegetable patches.

Community organizations

Another school worked with their local Rotary Club to develop a herb garden which used bricks from a demolished building to create spaces for planting, shaped in the form of the Rotary Club symbol. They gave their time and, of course, recycled materials to create a garden that helped children to learn about different herbs, how to grow them, their culinary and medical uses, and history. The garden also became a sensory garden with children loving the experience of being able to wander around and run their hands through the herbs so that they could enjoy their pungent aromas.

What will you do?

The obvious next step is to turn the wish list into reality with a plan of action, breaking a two-year period down into manageable and achievable steps, term by term, as the example in Table 2.2 shows.

You might like to begin by deciding what can be accomplished by teachers and children immediately, and allocate jobs and responsibilities to different people and year groups across the school. The next step will be to explore external links to find out which organizations and individuals could support the development of 'Science Beyond the Classroom Boundaries'. Do consider engaging children in this process by asking children to write to, and create reports and presentations to different external agencies, informing them about their ideas for their school grounds. You could challenge the children to think about how

Table 2.2 Action plan for the first year of changing the school outdoor environment

Immediate action	Science outdoor boxes	Hanging baskets	Tables outside each classroom	Outdoor display / photograph board showing personal skills in action
Term 1	Raised beds outside each classroom	Darwin's Thinking Walk	Bird box and webcam	Thinking posts – asking questions about 'How do you like to work?', 'What makes you a good team worker?'
Term 2	Wall murals	Hide and identification chart	Seating area – 'sitooteries'	Talking Tents – dedicated spaces to talk about science-related themes, introduced by teachers or by the children themselves
Term 3	Texture boards	Shade over workstations outside	Design and create a pond	Playground floor games related to working in a team

the person or organization could help them, focusing on what skills and knowledge people might have to offer to support different aspects, rather than asking for donations of money.

Finally, consider how you and the children will communicate all the exciting ideas and events to parents and friends of the school including school governors. This is particularly important in relation to parents, so that they understand the educational value of working in science 'Beyond the Classroom Boundaries' in terms of science skills and knowledge, as well as Personal Capabilities, which is the focus of the next chapter.

Practical ideas for teaching sound 'Beyond the Classroom Boundaries'

Sound is a topic that is best taught outside, for obvious reasons, such as:

- children can manipulate sound and make as much noise as they want, without disturbing other classes
- children have more space to carry out investigations such as exploring string telephones
- the space outside allows children to carry out sound investigations without sound interference from other groups
- children can explore concepts such as sound and distance more easily than in the limited space of the classroom
- the outdoors offers different materials for exploring sound
- the outdoors offers opportunities for using sound sensors.

One of the most exciting ways to work is to ask the children what they want to find out about sound in the school grounds, and also what they would like to create in the school grounds that make sounds. Then let them get on with it, with the teacher in a supporting role.

Here are some suggestions of using different kinds of science equipment outdoors.

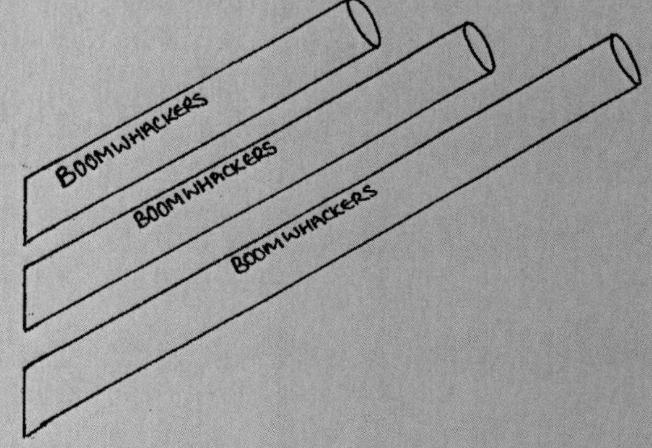

2.2 *Boomwhackers*

Boomwhackers can be purchased from primary science catalogues, but the can also be made from tubes of different lengths and materials – for example, plastic, card or metal (piping from a builder).

Children could investigate how the length, material and width of the cylinder affects the pitch of the sound, how to create louder and quieter sounds, or how to make a pipe band.

Hitting the pipes on different surfaces around the school grounds creates different sounds and these could be recorded using Easispeak microphones.

Children could use Easispeak microphones to record the sounds made and computer sound sensors to record how loud different sounds are.

Challenge children to explain how the sound of a Boomwhacker is made and changed. They could use photographs, Easispeak microphones or a video recorder.

Using simple thinking strategies for example, GRASP (Getting Results and Solving Problems) helps children structure their investigations. The GRASP framework leads children through different stages of thinking in a way that enhances their self- and group management skills.

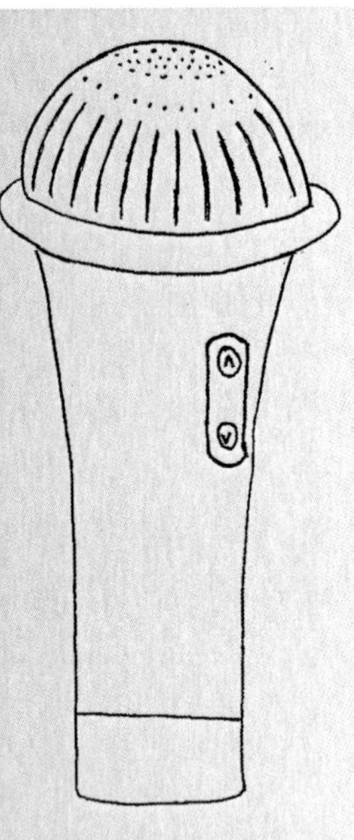

2.3 Easispeak microphone

- Step 1: What are we trying to achieve?
- Step 2: How will we know if we have achieved it?
- Step 3: What are the ways that would let us achieve it?
- Step 4: Which is the best way for us now?
- Step 5: How will we keep track of our progress?
- Step 6: How could we share what we learn?

Note: GRASP is a trademark of the Comino Foundation (a charitable trust operating in England). The wording of the GRASP questions have been adapted slightly for use here.

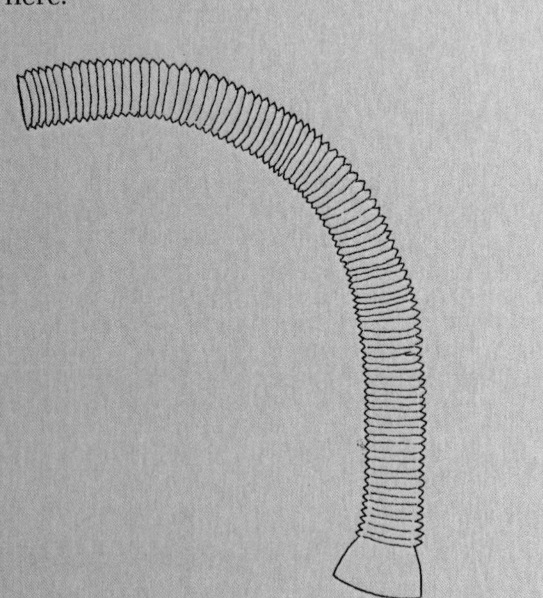

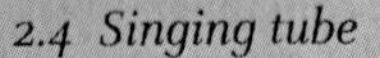

2.4 Singing tube

2.5 String telephone

Making and using string telephones are one of the many fun activities in sound topics, and offer lots of opportunities for children to explore and carry out fair test investigations. Show children how to make a string telephone and allow them time to explore it before challenging them to ask questions using different question stems (see Chapter 1).

how	what	where	which	who
why	when	what if	could	does
will	should	how does		

Once they have created questions they should then organize themselves to answer them, and decide which questions would be best answered by working Beyond the Classroom Boundaries.

There are many challenges that can be offered to children – for example:

- Who can create a record-breaking longest working telephone?
- How can you make a telephone system with more than two telephones? How will it work?
- Who can make a telephone that will work around corners?
- How can you set up a telephone system to allow people to make contact from the classroom with anyone working outdoors?

Using a broom handle or thick washing-line rope between two poles or posts, children can hang objects which can be struck to make a percussion area which can be explored by children when working or playing outside.

Children choose objects to add to the area which make different sounds when struck – for example:

- saucepans
- saucepan lids
- foil dishes
- metal tubing
- metal spoons
- bamboo pieces.

2.6 Outdoor percussion area

Children design and make their own wind chimes using recyclable items.

What kind of wind chime will they make? Ask children to pick a card that contains their challenge – for example, to make a wind chime that makes:

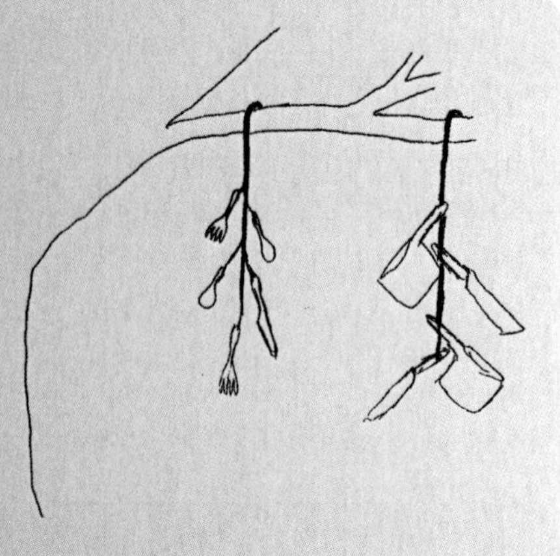

2.7 *Wind chimes*

- high notes
- low notes
- mixture of notes
- quiet sounds
- loud sounds
- tingling sounds, clunking sounds.

Challenge children to find out: which is the best place to site their wind chimes? What is the evidence? For example, have they carried out a movement test using sensors, or measured wind speed in different areas of the school grounds?

This experience allows for some ongoing tracking and monitoring. Children could choose where to site their wind chimes, but will only really understand whether it is a good place if they can see (and hear!) changes during different types of weather.

Keeping track and monitoring are essential skills of self-management, but they are difficult sometimes to set up. In this case, groups of children could take responsibility for tracking the evidence over time, during different weather types over the course of a week perhaps. Discussion could be had about how to effectively keep track – consistency of approach and efficient recording methods are crucial scientific skills too. The key to this is for the children to take ownership of the tracking and monitoring, where possible with as little prompting from the teacher! Reflective discussion when looking at the tracked data will tell a story of success in itself and allow for honest appraisal of how consistency in actions influenced the final results.

Working on sound provides an excellent opportunity for children to work with computer sensors, to measure sound levels. Challenge children to ask their own questions about different sounds and sound levels around the school. Begin by giving children question cards to extend their use of different question stems:

2.8 *Computer sound sensor*

how	what	where	which	who
why	when	what if	could	does
will	should	how does		

The answer to each question should be supported by evidence using ICT (e.g. sound sensors, Easispeak microphones) which allow children to record sound and download sound files onto a computer.

Question stems support different types of thinking from critical thinking, where children are using the questions to unpick and seek explanations and justifications for their observations, through to creative thinking where the question allow for free flow of open thought and exploratory reasoning. 'Asking questions' may initially seem simple – however, whether they are open or closed, the skill of asking a 'good' question is one that requires teaching, modelling and practice. This easy method, where self- and peer-review can be enhanced through recordings, provides an engaging and meaningful exercise for all children. The stems are a great way to scaffold this learning.

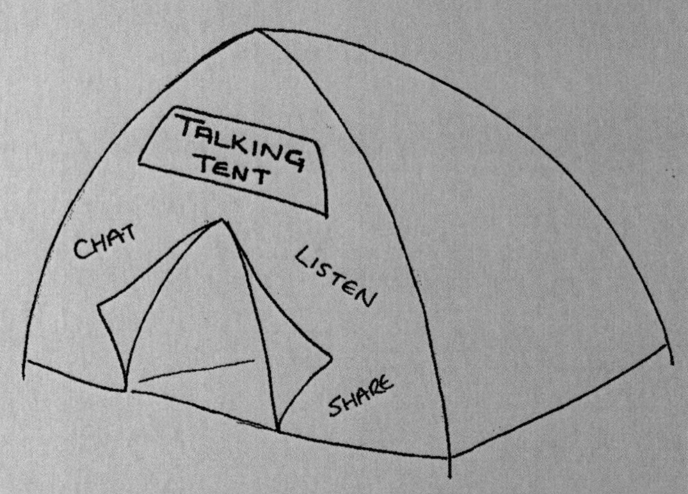

2.9 Talking Tent

Talking Tents are brilliant: they are simply a tent or covered area outside, which is dedicated to children going inside and talking to each other. Prompt cards with key questions, photographs, pictures and diagrams that are changed periodically can support this talk – however, the tent should be a space where anyone can talk about anything.

Easispeak microphones and string telephones could all be housed in these tents. Mirrors would also be great in here, just to encourage children to watch their own faces as they talk.

Different themes could be introduced into the tent, for example, talking with hands, talking without words, talking with faces (facial expressions), talking softly, talking with eyes and so on.

The key issue is to encourage the children to explore the range of ways we talk to each other and to find out the range of talk strategies people use in a fun way.

Using Darwin's Thinking Path (see Chapter 9)

There are a number of ways that the Thinking Path can be used to support a sound topic – for example, children could be challenged to think about how they would answer questions such as:

- how do the sounds along the Thinking Path change as we walk it – why?
- how do the sounds along the Thinking Path change over a day – or week?
- which is the quietest / noisiest part of the Thinking Path – how could we prove this?
- which would be the best place to put a wind chime, and what kind of sounds would suit the Thinking Path?

Celebration 'Sound Proms in the Grounds'

A lovely way to end a sound topic is to have a set of performances in the school grounds with children using sound makers and instruments that they have designed and made over the course of the topic. They could:

- tell a story or recite a poem with sound effects
- play their own compositions using their sound makers and instruments
- invite a band or orchestra to perform at their Prom in the Grounds
- invite parents to their Prom in the Grounds
- invite everyone to try the different sound makers around the school grounds.

Alongside this are many organizational issues that children could take charge of – such as invitations, advertising, an entry charge and refreshments.

Reference

Futurelab (2008) *Reimagining Outdoor Learning Spaces: Primary Capital, Co-design and Educational Transformation.* Bristol: Futurelab. www.warwickshire.gov.uk/biodiveristy (accessed 6 April 2010).

Personal Capabilities and 'Beyond the Classroom Boundaries'

What's the goal?

The link between the development of children's Personal Capabilities and exploring science 'Beyond the Classroom Boundaries' may not, initially, be transparent. In this chapter we suggest that when extending the range and breadth of learning styles 'Beyond the Classroom Boundaries' we must do it hand in hand with developing and supporting children's personal skills and capabilities.

If we are really out to create opportunities where children are increasingly self-directing, where they have options and choices to work inside or outside the classroom, in school corridors, playgrounds, libraries, halls and so on, then we will undoubtedly expect a sense of personal responsibility from each and every one of them. This is where the notion of developing children's Personal Capabilities of teamwork, communication, self-management, creativity and problem solving plays a key role.

The Smart Science materials (Bianchi and Barnett, 2006), developed by the Centre for Science Education at Sheffield Hallam University, explore how Personal Capabilities can be embedded in the science curriculum (Bianchi, 2002). The goal for us is clear; we can benefit children's learning in primary science (and also beyond) by underpinning their experiences and development of knowledge and understanding by tailoring learning to be as much about 'how' they learn as 'what they' learn. Children can become self-aware, responsive to their own strengths and after some time increasingly 'personally literate'.

Work undertaken at the Centre for Applied Positive Psychology (CAPP, initiated by P. A. Linley in Coventry) adds the dimension of personal development to this area, providing a strong rationale for developing children's strengths. The work of Govindji and Linley (2007) supports the strategies being promoted in this book for developing children's Personal Capabilities; CAPP described strengths to be, 'natural capacities that people yearn to use and that come from within'. By focusing and acting positively on people's strengths CAPP's research describes how people feel good about themselves, are better able to achieve things, and work better towards fulfilling personal potential. In relation to science 'Beyond the Classroom Boundaries', this provides a coherent rationale for focusing on the skills and capabilities that children are strongest at, while managing their weaknesses.

As you progress through this chapter and explore activities you will note how attention is paid to developing the language and underpinning knowledge and understanding for a range of skills and capabilities. This is crucial in developing Personal Capabilities, and therefore it follows that it is an important element of children working 'Beyond the Classroom Boundaries' in science.

The ultimate goal – personal literacy

Many frameworks have offered skills or competency sets that can be associated with developing children's personal responsibility and independent learning skills (see Lucas and Claxton, 2009). All undoubtedly have their worth; yet for the purpose of this book, five Personal Capabilities will be addressed:

1 Self-management – taking charge of your own learning.
2 Teamwork – working well in groups and teams.
3 Creativity – coming up with and sharing new or unusual ideas.
4 Problem solving – analysing problems and developing strategies and solutions.
5 Communication – speaking, listening and sharing feelings with meaning.

These capabilities have often been thought of as being endpoints in themselves – for example, once children are more aware and able to communicate, or manage themselves, they have achieved the required outcomes. Understandably, these first steps are the foundations of development, yet as with any form of construction the foundations lead to something greater, something which will be called being 'personally literate', a term synonymous with being an independent learner. Personal Literacy defines a range of abilities, from children having the knowledge and understanding of personal skills and capabilities, to being able to speak with confidence about their own personal development. Children who are developing in terms of Personal Literacy begin to articulate which particular aspects of a skill they feel is relevant to them and why. They also develop a strong sense of what they can do to help improve their abilities and many are able to think about where they could find support. Eventually, they have the self-confidence to demonstrate their skills to others; receiving and giving feedback where appropriate.

The list of skills provides a framework to stimulate a range of approaches that aim at achieving a number of success criteria; for example, children will have:

- developed an understanding of what it means to be a good self-manager, communicator, problem solver, creator and teamworker
- tried out and modelled a range of strategies related to the skills
- reviewed with others their skills, taking and giving feedback
- become more articulate at describing, explaining and exemplifying how the skills have affected their learning.

What is the reality?

Some national curricular and their accompanying assessment routines focus on learning defined subject knowledge and skills, while assessment is focused on how much children know. Strategies and initiatives that explore personal development have made inroads into appreciating the influence of our social and emotional skills on learning, yet often such experiences are not infused into regular subject learning throughout the course of a day. The greatest success in focusing on 'how' children learn, as opposed to only 'what' children learn, results from the teaching and learning that emphasizes the need for children to engage in discourse about their learning experiences and achievements with others. Self- and peer-assessment are becoming much more commonplace, if not routine in many classrooms, and the benefits are clear. In this book we take the view that the platform is now established to ask teachers to consider how such discussions about learning, with adults, peers and even with oneself, can focus on personal skills and capabilities. Much of the research done to develop these capabilities has shown that it improves children's self-awareness, raises self-esteem as children realize and respond to their successful learning skills and enables greater achievement as their pathways through learning the 'stuff' are better tailored to scaffolding 'how' they can be successful. As children move 'Beyond the Classroom Boundaries' in primary science the focus must be on developing children's personal capabilities.

Skills for Learning

A group of seven Sheffield schools involved in the research for this book worked as a 'Family', linked through a shared development called the 'Skills for Learning Framework'. This group, which includes a secondary link school, created a Directorate of Teaching and Learning, which explored how the skills such as independence, reflection and communication can be described, made explicit and encouraged in regular whole-school and classroom routines.

Across the schools the children are made aware of which skills they are developing in different activities and the schools share a common language relating to Personal Capabilities, which eventually is progressed into the secondary years. Displays, assemblies and rewards all serve to reinforce and celebrate the development of personal skills, and teachers use a wide range of active teaching and learning strategies to embed them in curriculum lessons. The schools are now engaged in extending this approach to 'Beyond the Classroom Boundaries'.

What are the options?

In this section we will share a model which describes some options for us to help children respond to and enhance their Personal Capabilities. As you read this section consider how this relates to developing children's ability to work 'Beyond the Classroom Boundaries'.

Figure 3.1 Model of Personal Capability development

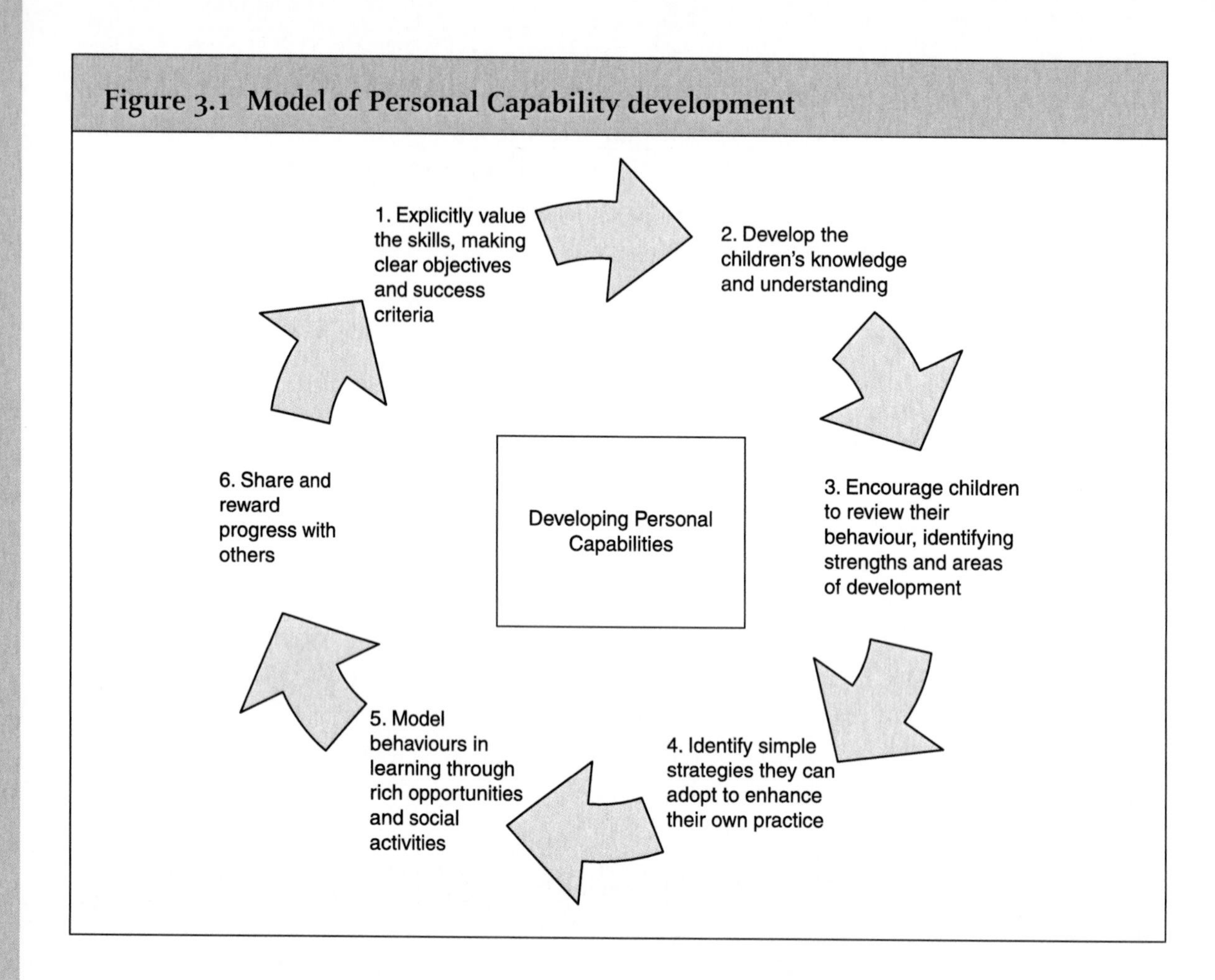

Step 1: Be explicit

Be explicit by making clear the objectives and success criteria; for example, how do they think they should behave outdoors? What should they do if they have a problem?

Making the capabilities explicit, both verbally and visually, are the first steps in describing them and placing value on the behaviours that underpin them. In Table 3.1 we have unpicked what is valued and shown what it might look like in relation to working 'Beyond the Classroom Boundaries' in science, which is based on material from *Smart Science* (Bianchi and Barnett, 2006).

Displaying these objectives on posters and boards outside in the school environment and sharing them with all members of the school community, including administrative staff, governors and parents, show the children that these things are valued.

Step 2: Underpinning knowledge and understanding

Develop the children's knowledge and understanding of a skill or capability; help them make sense of what the skills mean to them. Explore what types of behaviours, people and activities they associate with different skills and capabilities – for example, team-work. Ask: who in this school would you say works well in a team? Why? What does it

Table 3.1 Personal Capabilities and working 'Beyond the Classroom Boundaries'

Personal Capability	Examples to be able to:	Application 'Beyond the Classroom Boundaries'
Self-management	• Keep track of what I'm doing • Clarify what needs to be done • Organize and plan how to go about a task • Avoid giving up easily	• Be able to time-keep while working outdoors • Agree which task each person in the group will do • Work together to solve a problem – for example, how to measure a shadow
Teamwork	• Overcome challenges together • Make sense of what the team has done • Co-operate with others • Help reach agreements with others	• Return to the group to share results of what each person has done and consider if everything has been accomplished
Creativity and problem solving	• Consider why, how and what if? • Use different senses to stimulate ideas • Take time to be imaginative • Make decisions after exploring alternatives	• Use the Darwin Thinking Walk (see Chapter 9) to think through ideas and plan what to do
Communication	• Ask different types of questions • Share ideas, feelings and opinions with others • Justify opinions • Show ideas and information in different ways	• To take time before starting a task to talk through ideas and option • To discuss how they will record what they are doing when working outdoors

mean to be a team player? You will be amazed at how much the children already know about such skills, and possibly surprised about some misconceptions they may have.

As you begin to work 'Beyond the Classroom Boundaries' with children it will be necessary to introduce children to key ideas and language related to Personal Capabilities. At the same time children will also need to develop their understanding of how Personal Capabilities will help them to work 'Beyond the Classroom Boundaries'. Some teachers found it useful to use a range of scaffolding techniques to help to introduce and reinforce children's understanding of Personal Capabilities.

In this example we show how using stories and simple games to introduce a Personal Capability or skill can be engaging, relevant and fun.

Mr Men books

Some schools working with Personal Capabilities have been using cheap and cheerful books like the *Mr Men* and *Little Miss* series by Roger Hargreaves, which are great for this. Think about how Little Miss Scatterbrain can help children's understanding about what it means to be a self-manager; this is great for helping children to realize that it is important to be organized when going outside to work. What if Little Miss Scatterbrain went outside and forgot to take her science equipment?

Read children the story about Mr Bossy to introduce effective teamwork and the idea that children should decide who should take which role when they work outside and that to work successfully they need to listen to each other. From these discussions draw out behaviours that you and the children value and unpick these as much as necessary. For instance, if the children talk about how it is good to co-operate in a team when working outside, challenge them to describe what you would 'see' if they were co-operating – for example, listening to each other, taking turns. Even simple language like 'It's good to listen when communicating' can be broken down to actual behaviours that children can lock into, for example: what does it mean to listen well? What would I see, hear or feel if you were listening well? In taking this approach we are building up the success criteria on which children can base some reflective discussions. By clarifying understanding in this way children build a knowledge base relating to Personal Capabilities and what these look like when working away from the classroom; in other words, what kind of behaviour they should expect of themselves and each other when they are working 'Beyond the Classroom Boundaries'.

Once children are familiar with the language behind a skill or capability, highlight one or two objectives for the task. Sit these alongside the other objectives in your lesson – for instance, explain to the children that you will be looking to focus on the objective 'to be co-operative' while doing an investigation on living things in various habitats around the school. As such you will be providing them with double objectives: one focused on the process of learning (the Personal Capability) and one focused on the product (the science knowledge, understanding or skill).

Step 3: Self- and peer-review

Provide fun opportunities for children to think about their skills, what they are good at and what they think is good in others using self- and peer-review strategies.

There are many approaches, some as simple as verbal strategies, like 'Thumbs-up-sideways-or-down' approach,' (good, not sure, no) or 'Two stars and a wish' (two things that went well and one that I or we think we can improve on) can be used when reviewing activities against the Personal Capability and science objective.

The drive for using self- and peer-assessment has become greater since the launch of Assessment for Learning in the UK. It is important that when children are working 'Beyond the Classroom Boundaries' children are encouraged to think about how well they have worked, so that we celebrate and reinforce the development of Personal Capabilities.

Thumbs-up

Schools have taken the 'Thumbs-up' approach a step further by helping children to think about their thinking (metacognition) by challenging them to think beyond the initial judgement. For example, if children show a thumb up for having been co-operative when working outside (1st stage metacognition) they are asked to describe how they did that (2nd stage) and then pushed to explain why doing those particular things was helpful (3rd stage). Using peers as reflective partners also helps objectivity, and supports the celebration of positive behaviours. If you would like to find out more about this you can view some video clips on the AstraZeneca Science Teaching Trust (AZSTT) site by clicking on the Personal Capabilities CPD Unit (www.azteachscience.co.uk/resources/cpd/personal-capabilities/view-online.aspx).

Children can be asked to nominate someone who they felt displayed a Personal Capability; for example, co-operating well or justifying opinions. Of course, they are also asked to describe and explain their selection, drawing out the strategies the individual used and building confidence and self-esteem in the child(ren) who receive the nomination.

Step 4: Simple strategies towards success

Identify simple strategies that can help a child begin to use the skill or demonstrate it better. This crucial step, although often overlooked, acts as the lever to changing behaviour. Beneficial as it is to be knowledgeable about a skill and knowledgeable of how you currently demonstrate it is important in order to make a positive step towards improving or refining that skill or capability that one is aware and responds to simple strategies that can help move things on. Critical to any child's improvement will be their understanding of how to improve, what they can actually do on a day-to-day, hour-by-hour or minute-by-minute basis that will lead to a different outcome from what they have already experienced.

For example, when working 'Beyond the Classroom Boundaries' children might develop their understanding of the idea of effective communication and that it does not help if they consistently interrupt others, rather than stopping and listening. As children become aware that they often interrupt, they then need a mechanism or smart strategy to help them to change their behaviour. Help children to identify simple steps to take, for example, counting to 10 before sharing a comment, or making a note of a thought and saving it until there is an appropriate pause in the talk, would be clear methods to resist the interruption. The difficulty with this step is that there are no clearly defined answers to everyone's issues: no handbook to flick through as such. It is part of a teacher's responsibility, if engaging in the development of children's Personal Capabilities, to dedicate time to identifying, modelling and reviewing simple strategies that enable behaviour change. Time taken to do this in all learning but especially in relation to 'Working Beyond the Classroom Boundaries' will be time well spent. How else can children become independent thinkers and doers in science if we do not spend time and effort to support children.

3.1 Bums on Beds

Bums on Beds

Schools also use this fun strategy to support children's questioning in communication; it is know as '6 Bums on a Bed'. Although initially giggle-prompting this strategy focuses children on the question words that can help structure an outdoor investigation or exploration of something new or unusual. For instance, if children are considering habitats of living things, or the variety of materials that are used in the outdoors and why, children can be encouraged to use the 6 Bums on a Bed approach to tackling a question or dilemma – what, where, when, why, who and, of course, how. A simple yet effective strategy which helps children to make the most of their outdoor experience with responses that they and you will value.

Step 5: Rich, embedded tasks

Infusing or embedding skills and capabilities requires a proactive approach. We need to think of ways in which we can guide and facilitate children's learning during a task, so that what the children do addresses both the development of the skill, and the science subject knowledge. It is not enough for us to expect children to just know how to do the personal skills element of the activity. We need to help them to recognize the Personal Capabilities that they need to work on, such as identifying and taking on different roles in a group. When working in the more flexible outdoors environment we must be even more proactive in providing the structures to help them out, for example, thinking frames to encourage the gathering and refinement of ideas, or listening protocols to help each participant share their ideas, or time guides to assist them with managing a task to meet a deadline, when there is not a classroom clock because they are outdoors.

Embedding requires personal skills and capabilities to be incorporated in a rigorous way into the lesson objectives, success criteria and learning outcomes. Collaborative learning strategies lend themselves perfectly to developing Personal Capabilities and strategies such as 'Think–pair–share', the Jigsaw Approach, Shoulder and Face Partners are ideal ways to help children share their learning and its outcomes.

Using Outdoor Learning Partners

Teachers from project schools are beginning to adopt a range of different strategies when working 'Beyond the Classroom Boundaries'. This includes asking the children to identify an Outdoor Learning Partner (OLP). If you try this explain to the children that their OLP will be the person with whom they discuss the task, clarify what needs to be done, consider what they might need and come up with initial suggestions of how to do it. This could be a dedicated time for planning an investigation helping the children to be clear on their question, the method with which they shall explore and/or experiment and to be clear on the way they will record or evidence their outcomes. At this stage they could jot down for themselves on Post-it notes things they are sure about and things they are unsure about; they should collect these as a pair. When they have done this, they raise their hands to indicate to other pairs that they would like to share thoughts. They link up with another pair and share the things they are sure and unsure about – hoping to clarify as a foursome their remaining issues.

For a task to truly embed personal skills and capabilities you should be looking towards tasks which seamlessly join the personal and the science subject learning, where you can not actually do the science learning task without the skill nor the skill without the science. For more guidance and examples of embedded tasks see Smart Science (www.personalcapabilities.co.uk/smartscience/).

Step 6: Rewarding progress

The use of whole-school- and classroom-based rewards are useful in recognizing and endorsing positive behaviours with regard to Personal Capability development.

Award systems

Many schools have adopted award systems to support the development of Personal Capabilities in science and when working 'Beyond the Classsroom Boundaries'. Certificates of merit, Star of the Week Boards, stickers and Golden Time are just a very few of the many ideas schools regularly use to reward achievement, effort and learning of Personal Capabilities. Teachers encourage children to identify and explain who should be rewarded and why. As much as possible such rewards should reward effort and persistence towards developing oneself; after all not even adults are perfect at all personal skills, are they? For example, perhaps a group working 'Beyond the Classroom Boundaries', on finding out the quietest and noisiest areas in the school grounds, organized themselves to make sure that they took all their equipment outside with them, delegated each other a task to carry out, and used computer sensors to record the sound levels measured at each place.

The key thing to remember is to be positive, as with all learning: Personal Capabilities are strengths to support and enhance self esteem and self-image.

What will you do?

Life skills not classroom skills: Get as many people in the school involved as possible, emphasizing the same types of skill and using the same type of language. The more children see that there are consistent messages about these behaviours in and around the school the better. Remember that these should be shared with parents so that parents appreciate that they are as important as other forms of learning such as reading and writing.

Take and allow some risks: Challenge yourself to step outside the normal structured, teacher-led or highly scaffolded learning experiences, so that the children's skills and capabilities are challenged and stretched. This could relate to giving more choice or freedom with regard to who they learn with, what they learn and indeed where they learn! For example, when children are planning an investigation, include as part of that planning who they are going to work with, what they need to do and where they think it will be best to work, making sure that they understand that they can choose to work 'Beyond the Classroom Boundaries'. Preparing for these risks will help you to be more confident to take them, so talk about boundaries and strategies to show skills being used and of course sanctions where necessary.

Learner voice: Take learner views seriously; create opportunities for children to share stories and strategies, and seek ideas from other children or adults. It is essential that opinions are valued if children are to feel confident to engage with peer-review. Enhancing these opportunities will develop their confidence and trust in themselves that they are taking responsibility for their own development. For example, provide time for children to discuss with the rest of the class their ideas about working outside.

Active questioning: develop an atmosphere where learners ask questions of you and others about the way they can work or have worked. Ask them to question what challenges them now and how they will manage those challenges.

Catch confidence: Proactively develop ways to celebrate effort and progress with regard to Personal Capabilities by use of self-review and peer-feedback, as well as adult praise when they are successful.

Think smart, plan smart: Develop your own and children's sense of progression in Personal Capabilities, so that you and the children think about how within a few weeks, over a couple of terms or the school year, and so on, Personal Capabilities can be developed. Talk and plan as a staff how your school provision can be more responsive to children's personal development needs and requirements.

Practical ideas for teaching Personal Capabilities through science 'Beyond the Classroom Boundaries'

In order to help you picture what a science lesson embedded with Personal Capabilities may look, feel and sound like outdoors we have chosen an activity from the Smart Science pack that has proven most successful with its users. It focuses on the development of creative capabilities in focusing on the objective 'to ask why, how and what if?' while tackling the science investigative skills of 'making comparisons between pieces of evidence'. The activity name is 'Hedgehog Crime Scene'.

The generic task

The first steps in making clear the activity objectives and developing the children's knowledge and understanding begins with the 'If ... then ...' generic task. This game-like activity can take anywhere between 5 and 25 minutes and encourages the

Figure 3.2 Hedgehog Crime Scene Cards

Generic scenarios	Science-related scenarios
If trees could talk, then... If the stars came out just one in a hundred years, then... If pigs could fly, then... If promises had to be kept, then... If we were only one inch tall, then... If we had a TV remote control that worked on people then...	If people didn't have skin, then... If all green plants died off, then... If rocks were flexible, then... If people never had children, then... If all food was carbohydrate, then... If gravity didn't exist, then...

children to consider a range of 'if ... then ...' scenarios. Small cards with the scenarios are handed round or laid out as children sit in groups of up to six.

A child picks a card and completes the sentence. They are encouraged to consider more than one ending to the statement. During the game observe how much talking, smiling and laughing you hear. After the game has allowed each player a few goes, stop and ask the children what it felt like to complete the sentence.

3.2 Child with outstretched arms

The value of this generic and simple, low-cost activity is that the children begin to sense and feel what it means to think about a range of possibilities and to appreciate that there could be a range of viewpoints that are worth exploring. Giving them the 'feeling' of being creative and there being no rights or wrongs has been found to be liberating for a lot of children (and adults!) who play this game.

The embedded task

This task aims to take the creative questioning skills and to apply it to an engaging context in order that children can explore it again, while also developing their scientific investigative abilities of reasoning with the use of different types of evidence.

A great way to engage children in this task is to stage the hedgehog crime scene as shown in the following hand drawn scene in an outdoor setting (Illustration 3.3).

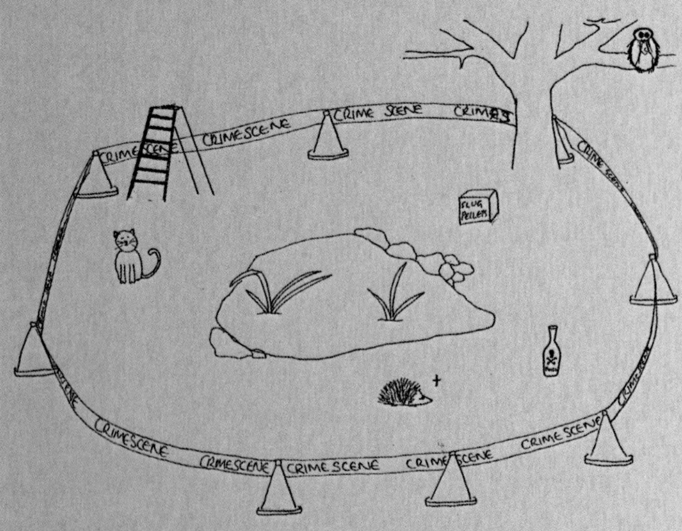

3.3 Hedgehog Crime Scene Cartoon

With the use of puppets and a few props such as a bicycle, wheelbarrow and pretend Slug Pellet and Weed Killer boxes the scene is set, made even more authentic if surrounded by crime scene tape (easily purchased online).

The children's task is to ask 'why?', 'how' and 'what if?' questions in relation to the scene, where they can see an unconscious hedgehog lying

Figure 3.3 Hedgehog Crime Scene

forlorn on the ground. Teacher in project schools have taken on the roles of Mr and Mrs Hog (the hedgehog's parents) to further investigate the circumstances surrounding the scene. Initially, children are encouraged to be open-minded and to brainstorm a wide range of possibilities. As further evidence can be introduced they are required to select those ideas that may be more plausible. Using a simple thinking framework they can be supported to arrive at a definitive thought or couple of thoughts, which they describe and justify using the evidence provided.

Discussion can be had with regard to what types of evidence are available, which are more reliable than others, and which could be supported by further investigation. If desired investigation plans can be outlined and even undertaken; for example, exploring if the hedgehog had consumed slug pellets which made him ill, children could be given or calculate the average number of slug pellets per box and compare it against the box found at the scene.

The use of self- and peer-assessment strategies; for example, 'Nominate Someone Who', involves the children in reflecting not just on their own contributions, but those of others, and to identify and explain the personal skills and capabilities that have been used, and the feelings, strategies and types of discussion that let them flourish.

Hedgehog Crime Scene in school

Wooley Wood Special School also worked on the crime scene with severely disabled children, by using their senses to fully explore all the aspects of the scene, from the grassy verges, to the spikes of a hedgehog. They collected garden matter from the outdoors and created a very large frieze of the scene; they invited specialists to the school who brought hedgehogs in and allowed the children to touch and smell the animals; and they explored the feeling and movement of water and the smell of burning leaves on a fire. A wonderful approach that led these special children to engage with this task and gain relevant learning for their abilities and interests.

References

Bianchi, L. (2002) Teachers' experiences of the teaching of personal capabilities through the science curriculum, Ph.D. thesis, Sheffield University.

Bianchi, L. and Barnett, R. (2006) *Smart Science: Activating Personal Capabilities*. Sheffield: Centre for Science Education, Sheffield Hallam University. Available online at www.smart.science.co.uk

Govindji, R. and Linley, P. A. (2007) Strengths use, self-concordance and well-being: implications for strengths coaching and coaching psychologists, *International Coaching Psychology Review* 2(2): 143–53.

Lucas, B. and Claxton, G. (2009) *Wider Skills for Learning*. London: NESTA.

CHAPTER 4

Managing children working 'Beyond the Classroom Boundaries'

What is the goal?

Underpinning Science 'Beyond the Classroom Boundaries' is the idea of developing children as independent learners. In the context of working outside, our goal is to create an atmosphere and range of situations where 'Children can gradually move from regulation by others to self-regulation if appropriate frameworks and strategies for eliciting and acting upon their views are put in place' (Futurelab, 2008: 24).

Our aim is to encourage children to act and think for themselves, promoting confidence and a range of essential skills and capabilities regardless of the variety and range of contexts, abilities, settings, teacher confidence or pupil independence. In this chapter we share some of the ways in which schools have overcome their issues to facilitate this kind of practice so that teachers and children can work successfully 'Beyond the Classroom Boundaries' in science.

What is the reality?

As we have highlighted, the reality of science in many schools is that as children move through the primary years there are fewer opportunities for children to work outside, and even fewer to do so independently.

As part of our work with schools we asked teachers what stopped them from working with children on a regular basis in science 'Beyond the Classroom Boundaries'. Here are just a few of the comments we have collected from teachers; perhaps you recognize some of the issues as your own:

Can't always see children – are they on task?

Weather – we've got a boggy field!

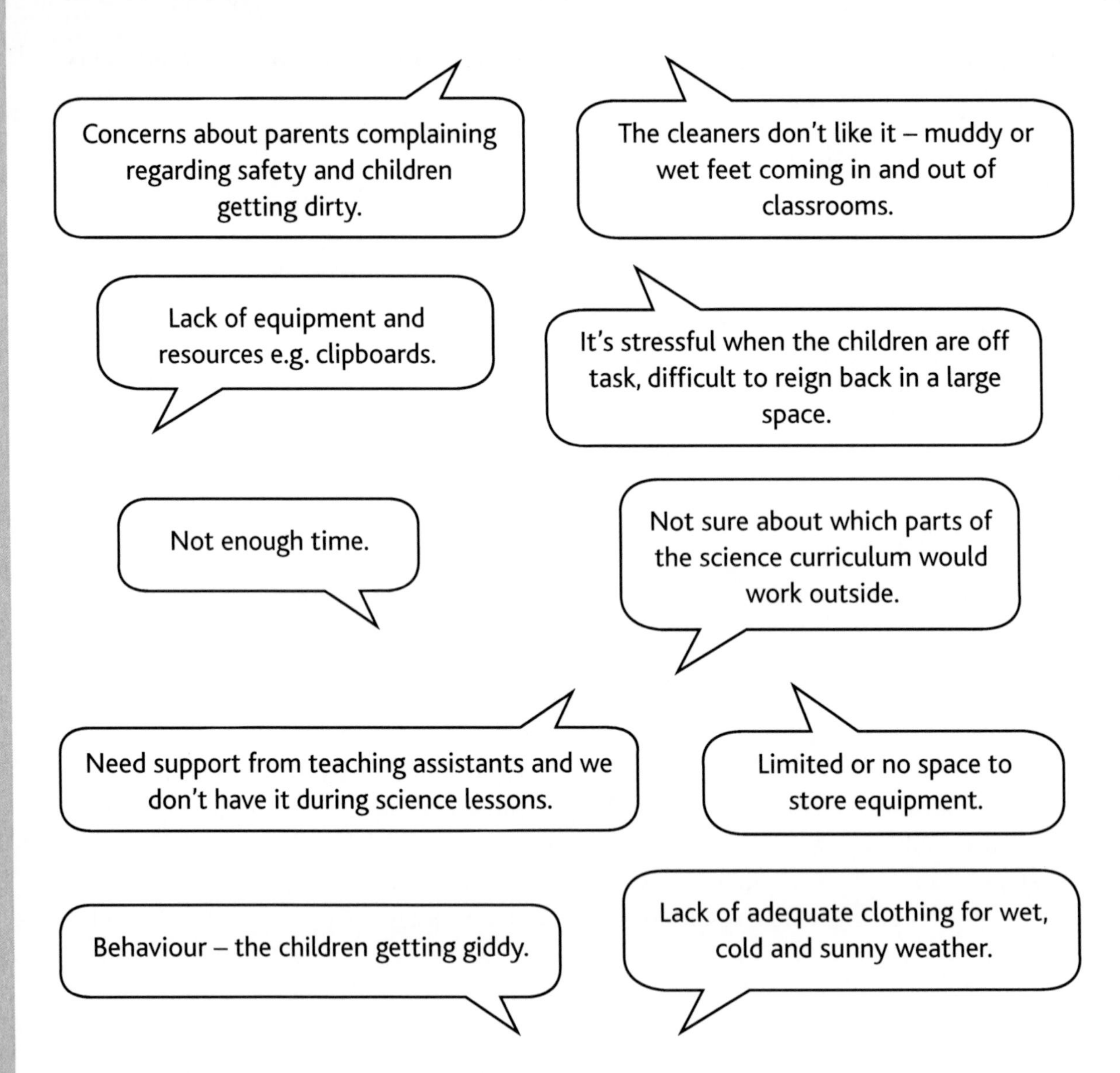

When thinking about moving science 'Beyond the Classroom Boundaries', we need to think about two types of space: *our head space*; getting the children and teachers in the right frame of mind, with the confidence to go for it; and the *physical space*: organizing areas or zones within the school which are safe places for children to think, do and learn in science. This may also require a change in mindset so that everyone in school understands that children in these spaces are engaged in planned science activity.

Sorting out the head space

Of course the first thing we must do is sort out 'our head space' so that all staff in the school have a shared understanding of the potential and purpose of moving 'Beyond the Classroom Boundaries' in science, so that it is the norm, not the exception. The teachers in projects schools made this initial decision about children's access 'Beyond the Classroom Boundaries' for science and worked to ensure that staff across the school developed a shared understanding and philosophy in science and worked together to move practice.

The reality is that in order to achieve the goal of children working as independent learners outside the classroom, we have got to decide whether we are going to be put off by the reasons why we think we cannot, or should not, take learning outdoors or, whether in fact, we manage those issues in a way that makes them non-issues. To be able to take this option of extending the idea of the classroom beyond its physical barriers, we need to develop strategies to manage key areas of concern; we want to feel confident that the children:

- are safe
- are able to cope
- understand what we are looking to achieve and how.

In the following section we look at some of the barriers that teachers have shared with us and share strategies from the project schools to remove them.

What are the options?

Weather, dirt and parents

This was invariably the first set of barriers that teachers put forward when asked what prevented them from working outside in science.

Rain rain go away,
Come again another day!

Would that it were so easy! Well actually it is. If early years settings can get their head around the issue of weather with children who find it difficult to do up their own coats and shoelaces, then surely it should not pose a problem for older children. It is all about asking the question 'Will learning Beyond the Classroom Boundaries elevate my teaching approach and the children's learning potential today?' If the answer is positive, then don the kagoul, put on the wellies and off you go. Interestingly, teachers in the project schools admitted that often the only reason for not going beyond the classroom was that it was cold or a little wet for their liking, the children were not bothered! We adults are often the ones who limit the scope for children's learning and once we acknowledge this ways of overcoming the weather are obvious.

Nina, Jess and Pauline at Grenoside Primary realized that among the staff there was a mindset that says 'we will do more when the weather gets better'. So they decided to tackle this by sending out a letter to parents asking them if they had wellies and waterproof coats that could be donated and reused in school, so that all children could go outside to work in wet, cold or windy weather.

They realized that wellies and waterproofs available everyday are the norm in early years settings, so really this approach should continue into later primary years.

They even decided which parent would help to make a 'wellie stand' to keep them tidy. Problem solved!

Interestingly, once teachers had realized that the answer to the weather was to kit the children out (and themselves!), they were determined to sort out this issue. In one school where funding this solution was an issue they decided that money could be raised through parent–teacher activities and through fund-raising, such as 'Wear your own clothes to school days' and having 'Science Fun Days'.

As for parents, they are an important partner in children working 'Beyond the Classroom Boundaries'. Teachers from project schools who have begun to work frequently outdoors in science took the decision to inform parents of the changes taking place in relation to teaching and learning in science. For example, some schools:

- sent a letter home to parents explaining why and how children would be working in science
- created a leaflet for parents about 'Science Beyond the Classroom Boundaries' and Personal Capabilities
- organized outdoors workshops so parents could experience for themselves the value of working outdoors in science
- asked for parent helpers to work with children 'Beyond the Classroom Boundaries' in science lessons.

Managing children's behaviour

Of course, we must be realistic and acknowledge that just opening the door and letting the children out is not what 'learning effectively' outdoors looks like. 'How can children be managed/supported in order to work safely when learning outside of the classroom?' was the question most frequently raised by teachers and undoubtedly a crucial one for all of us to consider. Concerns are frequently expressed relating to behaviour management and the worry that if children work outside they are more likely to be less well behaved, whether the teacher is with them outside or not. This was a very real worry for many teachers involved with the project but the responses were both interesting and heart-warming as teachers began to realize that it was less difficult than they had anticipated and that, with appropriate management, children responded very positively.

Good old basic classroom management outside

Carol at Shaw Primary reflected on her thinking as she began to use the school grounds more frequently in science, she explained:

'What I realised was that when you think about it, managing a class outdoors is not really much different than managing the children indoors. When we work in the classroom we have a shared understanding about rules and routines, all that we need to do is to make sure that children understand that these also apply when they work "Outside the Classroom Boundaries".'

'It seems obvious now, but funnily enough the first time the class went outside to work they flew out of the room they were so excited and would you believe I had explained the activity but forgotten to give them instructions about behaviour when working outdoors.'

Lesson learned! The second time the children went out rules were in place, the children understood the physical parameters and they were fine. Just good old basic classroom management applied to the classroom outdoors.

Surprises from special needs children

A teacher from Woolley Special School describes the effect of working outside on children with a range of special needs.

'In the classroom our special needs children can often be very challenging in their behaviour and as a consequence many staff were reluctant to move towards teaching the science curriculum using the outdoors. The great thing was that despite these concerns staff did have a go and what a surprise it was: rather than behaviour becoming more challenging, the children seemed to be much calmer outside than when in the classroom. In the classroom they would often become agitated but outside the children were more relaxed and focused when working in the school grounds.

Not only were the children more calm they also took the initiative, which included clearing up after themselves, something they certainly were less likely to do when working inside the classroom.

We also noticed that there was great difference in the level of co-operation between children: they worked together, sharing and helping each other.

It would seem that when the children are offered very real experiences outside the classroom boundaries, where their learning has a real purpose, teaching and learning takes on a different character.

The staff reflected on this, and wondered whether it was the perceived physical freedom the children experience, learning away from the tight confines of the classroom. Perhaps it was also the novelty of being outside when learning, where there are many things to stimulate the children's interest. Could it have been that the staff were physically less close to the children so that children felt that they could be more independent, or was it, as one teacher suggested that just being outside has a calming effect on the children – being with nature. Many children with special needs do not get outside much since some homes do not have a garden, and some parents find disabled children challenging to take out and so don't very often.

What we now appreciate is that all children should have access to working outdoors on a regular basis and this experience has given staff the confidence to work "Beyond the Classroom Boundaries" in science on a regular basis.'

Teachers from project schools found that it was a matter of setting some clear boundaries: establishing positive working practices, and taking it on the chin when something went a little off track in the early stages of working outdoors.

Children understanding their own Personal Capabilities

Across the schools there was a realization that central to working in science 'Beyond the Classroom Boundaries' was the requirement to underpin all work with the use and development of Personal Capabilities. So, in parallel to the 'teacher's head space', we need to think about 'children's head space'. This requires teachers to help children develop an understanding that they will be spending increasing amounts of time outside the classroom working in science and that there are expectations in relation to behaviour and ways of working. Their Personal Capabilities will be fundamental to being in the right 'head space' when they go out to work. You will need to encourage them to think about how they learn and to be conscious of what challenges them. This will mean that the teacher will need to talk to the children about working outside and the strategies that help them cope and keep self-control.

Teachers will need to establish ground rules for learning outside, which means reminding children about established school and classroom routines, and that these do not necessarily change just because they are outdoors. Knowing it is against the rules to climb trees, because it may lead to a serious injury, and the sanctions if this rule is breached is essential information. However, children require more than just a set of rules; they need to know how to work effectively in groups or teams, how to persevere with a task if it gets a little tough, how to share what they have learnt, what they can do to sort out an issue or problem before running back to the teacher, how to keep track of and manage jobs and time, and other vital learning skills. In a sense it is the skills of managing learning that they need in order to be successful when they engage in science outside the confines of the classroom. The questions that we need to ask relate to their Personal Capabilities, for example:

- Do the children know how to organize themselves to achieve the right things in the time given?
- Do they know how to sort out what they find or observe, so that notes or scribbles mean something when they get to share it with others?
- How do they cope when things get tough and they have a problem?

Sorting out the physical space

Once the decision has been made about working in science 'Beyond the Classroom Boundaries', the next set of questions relate to how the rhetoric translates into reality.

In many schools the decision is made by the school building. It is easy for those teachers in classrooms with a door which leads directly onto the school grounds to make the most of that immediate access to the outdoors and develop an 'open-door' policy similar to that found in early years settings.

In some schools, staff were happy to allow children access to any part of the school grounds. For others extending the parameters was a more cautious move with clear decisions where children were allowed to go and what level of supervision they required. All decisions should be based on where learning outcomes can best be met for the children, then the confidence of the individual teacher will determine when and where they feel they are happy for children to work outside. Creating 'safe zones' can be simply achieved by coning off particular spaces, where children are free to work, explore, pick things up, sit around and talk, think and work in science.

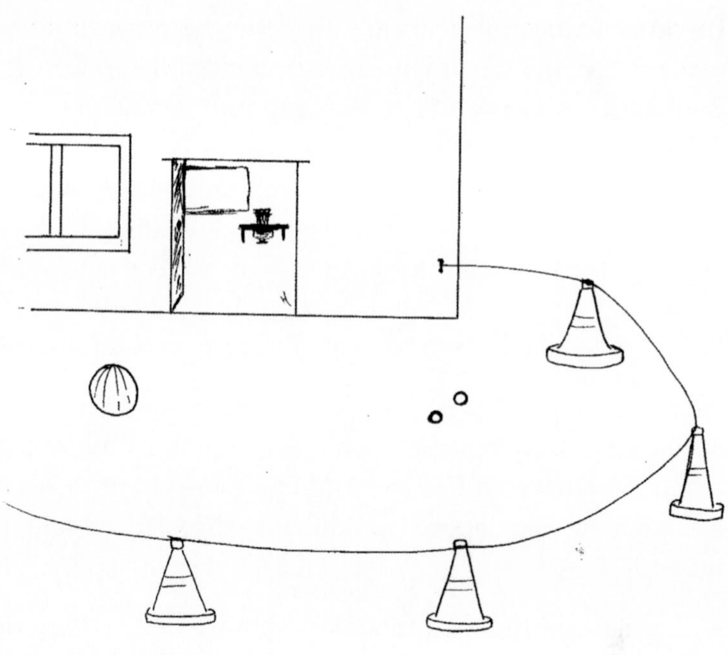

4.1 Coned areas

Any parent or carer will be familiar with doing some housework or cooking the dinner while their little ones are out on the decking or playing in the lounge and the level of risk is minimal, because those spaces are set up appropriately. As children mature, not necessarily by age but in responsibility, the safe zones should be modified accordingly.

In these areas it is not necessary to have constant adult supervision if appropriate ways of working are set up. For example, some teachers coned off an area close to the classroom so that children could work outside carrying out fair tests, or collecting invertebrates or using water to explore forces. In this scenario the teacher can easily see from the classroom what is going on and does not need to be in attendance outside. In other classrooms more confident teachers allowed children access to areas further away from the classroom: some still in sight of the teacher, others not so, but all set up in a way that the teacher and children were comfortable with the parameters and able to work well within the designated area.

Where teachers did not have immediate access to the outside their decisions were slightly different but again depended on teacher confidence and children's Personal Capabilities to work outdoors with limited or no direct teacher supervision.

- Will I need to take the whole class outside? If so how will I manage the class?
- Do I need to use additional adult support to go out and support a group of children working 'Beyond the Classroom Boundaries' in science? Do I do this in, for example, PE?
- Are children able to go out in small groups to work in a defined safe area without adult supervision?

Understandably, there are issues around manageability and in a few schools the most realistic options for providing learning opportunities in science 'Beyond the Classroom Boundaries' may need to be very carefully organized.

Our ultimate ambition though should be for children to be able to make decisions about where to carry out their science and be able to access the outdoors when they feel it appropriate to do so. There are many different strategies and most are very simple to apply. In the following section we share some examples from project schools.

Children deciding when to work 'Beyond the Classroom Boundaries'

All these approaches help children to develop a range of Personal Capabilities and those schools that take this approach make sure that children are partners in making decisions about how they will work outdoors. One thing is for certain though when trying to establish any new approach and related routines: patience is a virtue and consistency a necessity!

- Engage children in discussion about whether they need to work outside, when and how during the topic.
- Create a rota that indicates which groups have access to the outdoors and when.
- Allow x number of children out for a set time: give them a stop watch or sand timer.
- Children log out and log in using a white board, with a maximum number of children allowed out at any one time placed on the board.

The adult needs to scaffold children into roles where individually and in groups they can manage their own learning. This may take time but is important; it can begin with something as simple as putting children into 'role' through using props such as *Science Haversacks*, Kit Bags and Role Badges (see Chapter 7 for more information on the Science Haversack). A number of teachers found this a very useful approach; children began to think differently about going outside, realizing that they needed to be prepared for when they were going outside the classroom to work in science.

Teachers extended the use of the haversack or bag, tailoring it to support and develop a range of Personal Capabilities such as time management by including:

- timers which beep when the children are due to return to the classroom or allocated place
- stopwatches/watches

Badges are frequently used in classrooms to help children organize jobs in groups. They provide an ideal prompt for a child to take responsibility for an aspect of work, and to become aware of the different aspects related to the learning taking place.

Different kinds of activity outside the classroom might require children to think about the different tasks they must carry out to complete their work. For example, if children have decided to go outside to survey what kind of invertebrates are found in the school grounds, they might decide to organize themselves so that they have:

- *Timekeeper* – who thinks about 'How long have we got?' 'How much time do we have left?'
- *Recorder* – who is in charge of recording where invertebrates are found.
- *Communicator* – who leads the group to think about how to present what the group has found to the rest of the class.
- *Task manager* – who organizes what the group does, the equipment and helps them to think about if they have done everything they need to do.

4.2 Personal Capabilities Role Badges

In a different kind of activity such as a fair test investigation the badges might include:

- *Chief science resource manager* – What resources do we need? Are we using them correctly? Do we have all the equipment now that we are going inside?
- *Chief science fair tester* – Is our test fair?
- *Chief science measurer* – What will we measure? What will we use? Are we measuring correctly?
- *Chief science recorder* – How will we record what we did and our results?

4.3 Science Role Badges

Children managing their own problems

Some of the skills children need to manage their work outdoors effectively can be addressed in this way. However, the affective or motivational aspects of learning during a task are more difficult to scaffold as they rely on personal commitment to learning. Whether a child avoids giving up easily or takes time to be imaginative are aspects of learning that generally only come about when people sense they are valued behaviours for which they have felt or seen the benefits in the past. For these capabilities visual cues, posters, recognition and rewards are initial steps to remind children of the worthiness of a particular capability and its value in learning. As such, badges or visual cues are all ways of making real and explicit the skills that lay the ground for successful working beyond the classroom.

You might like to consider the idea of Thinking Posts around the school to encourage children to think about their Personal Capabilities. These remind children about capabilities that are valued and also celebrate them; for example.

- 'When we work outside we help each other.'
- 'When we work outside we try hard to solve our own problems.'
- 'When we work outside we don't give up, we persevere.'

As with all learning, and that includes children working on their own Personal Capabilities, it is important to value and celebrate children's achievement. So when children have completed their tasks 'Beyond the Classroom Boundaries' do encourage children to talk about not only what they did and found out in relation to science but also about how they worked, which Personal Capabilities they used, improved on and those that they think still needs some work.

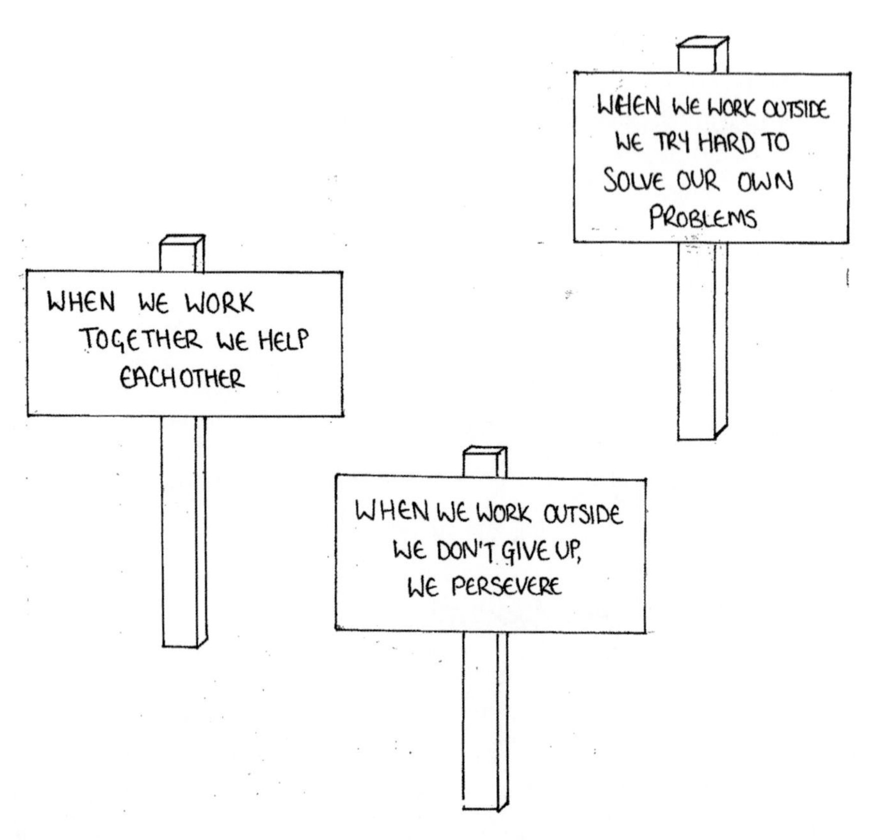

4.4 Thinking Posts

Communicating with children working 'Beyond the Classroom Boundaries'

As a class begins to move within the new learning parameters 'Beyond the Classroom Boundaries', both teachers and children might need the security of keeping in contact with each other. Project schools found interesting and imaginative ways of doing this, for example:

- *Walkie talkies* were chosen by some teachers who decided that the children would really enjoy using them, and that when children were working independently and out of sight of the teacher, this would be an excellent way of communicating and also support the extension and review of learning during tasks.
- *Mobile phones* were also a suggestion: a simple pay as you go with only one number which is the mobile back with the teacher. Children and teacher use the mobile to send text messages or have conversations relating to what the children were doing, how they were managing, and so on.
- *Science Bibs* have been very popular with project schools; some used PE bibs, while others purchased special Science Bibs with planets on. This provides a visual communication which tells any adult in the school that the children wearing these are carrying out science tasks 'Beyond the Classroom Boundaries.' Interestingly, children are proud to wear them, and also realize that they can be challenged if they are not working appropriately by any member of staff; conversely, schools using this approach also agreed that staff should compliment children seen working well outdoors. The Bibs help everyone feel secure that children are not just wandering around the school and as a relatively cheap and cheerful resource these transmit key messages of 'I'm learning', 'My teacher is aware I'm out of the classroom.' 'Please do not disturb – job in progress.' 'Do come and ask and we will explain what we are doing.'

Checking that children's learning is on course

Children being 'in charge' of their own learning is great, but only if we are sure that the learning is authentic, on track and purposeful. During the research for this book we found that teachers quickly identified areas of the curriculum and activities that could move outdoors. Often such activities had detailed planning and learning outcomes that were clear for the pupils. Project teachers found that the issues they faced when the children were working outdoors were the same as when children were working indoors. In the classroom teachers needed to check that children understood the activity and concept and the same was required when children were working 'Beyond the Classroom Boundaries'.

Catching mistakes early

While one group of Year 5 pupils in Sheffield were undertaking a group activity on Earth and Space outdoors, the teacher and other pupils remained indoors undertaking other themed science tasks. The children were using their science and numeracy skills to plot out, in proportion, the distances between the planets, across the playground.

Despite the group working out the correct distances from the Sun, they were inaccurate in where to start measuring from. Instead of measuring from the Sun each time, the children ploughed on measuring the distances from one planet to another.

On reflection the teacher described how she was disappointed not to have 'caught' the mistake at an earlier stage, and felt that the children had not achieved the learning outcomes due to one repeated error. She thought that this was a disadvantage of children working on their own outside and wondered whether she should only give them easy activities for outdoors.

We talked with her a lot about the need for children to be given appropriate challenges to stretch their learning while outdoors, and to avoid the temptation to give them easy tasks. After some discussion it was decided that together with a challenging task comes the need for support and direction at key points; the teacher decided to call them 'milestones' in the learning process.

These 'milestones' were a point in the activity when children needed to 'check-in' with the teacher to enable the children to explain what they were doing and the teacher to 'check' they were on the right lines.

These are very easy to set up, with simple statements, such as 'When you get to that point, please "radio in" so we can have a conversation, or "come back" so that we can check progress before you go any further.'

The teacher was able to overcome this issue and ensure that she provided enough challenge to the children to extend their learning and motivate them, while providing key markers to allow them to self-monitor progress.

Off task

A teacher from Bradway School, Sheffield, had a different problem and that was when working with the whole class outside one day, she saw that one of the groups was for some reason off task. She decided to call them back to her and as she did the whole class came back!

Working with this teacher about different strategies for managing individual groups outdoors, we talked about each group possibly having a number and the teacher could call/ring a bell and hold up the group number card that she wished to come back to her. This would allow the other groups to continue while she worked with one particular group.

The role of the teacher

From task setter to facilitator, from questioner to instructor, what is the teacher's role when encouraging autonomous learning outside? Where should they stand? Should teachers locate themselves in one position or move around from group to group or work with one particular group?

These are very real concerns for teachers when managing a class of children and they are no different to those asked when teaching science in the classroom. The only difference is the distance: in the classroom the children are nearby, in the outdoors the children might be further away from the teacher.

Probably the best way for you to find the answer to how your role changes is through trying out some science outdoors. In some schools the teacher and children were supported by teaching assistants or other adults and so the teacher could work with specific groups. Other teachers who did not have such support found they felt the need to maintain a position where they could have a watchful eye over all pupil groups, which also meant that the children knew where the teacher was. In some instances, for example, working with children on the Solar System or how to use a piece of science equipment, learning was more through teacher demonstration activities outdoors. There were other occasions where the teacher remained within the classroom while leaving the external door open so that some groups could work outdoors. Much was dependent on the learning outcomes, and indeed the physical locations of people's classrooms to open spaces.

The answers to what a teacher's role is are as numerous as the ways in which the teacher might work. What is obvious is that you need to explore roles in which you can feel in control of the teaching and learning, taking place outdoors as much as you do in the classroom. Your practice in terms of stopping and reviewing lessons should not change that much from indoors. Your role to challenge, prompt questions and extend learning is just the same – only the location is different. Do not feel that because the children are outdoors it is not possible to achieve high-quality learning outcomes that you have planned for: it is not a free for all unless you wish it to be! Structure and scaffold the learning, as well as the children's personal skills and capabilities, through verbal and written instruction, clear aims and success criteria, with time markers and responsibilities clearly set out for all to understand.

What will you do?

Teachers from project schools found that the best way to work was as a whole staff, so that eventually a whole-school way of working emerged. The starting point was invariably to provide a forum where staff could raise concerns and issues about working 'Beyond the Classroom Boundaries' in science. Once these were acknowledged, the next stage was to challenge staff to come up with positive ways round those concerns, and as the example below from Castleside illustrates, none of their perceived problems were insurmountable.

Remember that it is as important to prepare the children as it is for you to have planned the spaces and experiences for them. Essential to this is a clear understanding of your expectations for learning outdoors. Share your thinking with pupil groups: for example, School Council, Outdoor Pupil Action Groups and even parents beforehand. Model positive behaviours and practices; start with shorter focused activities and work towards more open-ended, team-based challenges. By reflecting on learning as well as behaviours with the children, using basic Assessment for Learning (AfL) strategies, self-, peer- and group-reflection, your expectations and their views on learning can be openly shared, celebrated and issues mitigated.

Table 4.1 Issues and strategies for working 'Beyond the Classroom Boundaries'

Issue	Suggested strategies
Health and safety	• Behaviour guidelines. • Sanctions for inappropriate behaviour. • Children work in pairs or more, learning to look after each other. • If there is an accident (e.g. child falls over, partner leaves child if necessary and tells adult in school). *Foundation and Key Stage 1* – issues are limited because the area is enclosed where they might work. Where the children use the wider school grounds then the additional staff are available. *Years 3 and 4* – have decided that the children will be given: • Oral boundaries • Timer • Badges with one denoting who the team leader is, this child is in charge of the group and reports incidents to the teacher. *Year 5 and 6* • Timers • Badges – with team leader badge
Time	Time is only an issue if this is seen as an 'add on' to science lessons rather than a regular and integral part of science. Time is an issue if children have to change clothing (e.g. shoes and coats), but with older children this should be minimal as they are more capable of managing themselves, which is in fact what this approach advocates – children being independent.
Ratio of staff to children	We realize that this is not an issue in the Foundation Stage because of the adult/pupil ratios. In other year groups this is dependent on the confidence and ability of different groups and the teacher. In some of the older year groups the children are in an enclosed area – the school grounds and the teacher can limit the area children use if the teacher lacks confidence in the children. The use of the school grounds should be seen as the classroom without a roof – the fencing provides a boundary. Class rules for working in the outdoors should help the issue of needing additional staff, as children become able to 'police' themselves, staying within a given area and behaving responsibilities – thus developing SMART Personal Capabilities.

Issue	Suggested strategies
Control	As above – usually children are so highly motivated that behaviour is not an issue. Class negotiated rules, giving children responsibility badges, time limits outside (children take a timer/ stopwatch out with them) all of these support strategies help to support positive behaviour. Children at Castleside are familiar with being given responsibilities and work positively, so this will just be an extension of the school expectations and pupils' expectations of themselves.
Weather	• Foundation have wet weather clothing • Other year groups could develop this system • Children put coats and wellies on to go outside • Cold weather should not be a deterrent • Summer weather – hat and sunscreen if out for extended periods
Resources	• Some purchases will be necessary • Existing science resources should be used • Local secondary partner school might offer support and loan equipment • Friends of school (or equivalent) might fund-raise • If everyday items such as seed trays in saucepans for a sound washing line, a letter to parents for donations would be appropriate
Differentiation	Should be based on approaches used in the classroom and appropriate to the activity, e.g. levelled groups for fair test investigations, mixed ability when they are exploring or carrying out surveys, sometimes groups will have reader with non-reader.
Confidence	• Teacher confidence – support from senior management, science leader • Child confidence – developed based on teacher showing confidence in them and having high expectations. As children work outdoors on a regular basis in science this will improve.
Keeping children on task	Hopefully motivation in relation to being allowed the independence to go outside will help to keep children on task, but it will help children to be given a stopwatch/timer, so that they know when their time period is up; this will also help them to develop the ability to time-manage when outside.
Which areas of science?	Only areas that are appropriate: for example, it is not really appropriate to make circuits in electricity outside; the pieces are small and easily lost and there is no practical advantage to being outside, unlike, for example, topics such as sound and forces.

(Continued overleaf)

Table 4.1 Continued

Issue	Suggested strategies
Do all children need to have the same experience?	No, different children can be outside for different reasons; what is important is that all children have regular access to the outdoors. In some cases (e.g. use of sensors), it would seem appropriate that all children experience using sound sensors outdoors, but it might be at different times of the year and in different learning contexts.
Fair testing	Where children fair test outdoors, for example, cars down a ramp, they will be thinking about what is fair, but perhaps the different surfaces are not even, then the children have to decide if the difference is so great it will affect the results and think about an alternative – this is excellent problem solving.

Practical ideas for teaching materials 'Beyond the Classroom Boundaries'

Testing materials

In this section we consider how a selection of activities related to materials can be completed outdoors and what strategies the teacher can incorporate to support children in managing themselves when working outside.

More often than not children find it easier and more motivating when science is placed in real contexts and children are given a problem to solve. Here the context for the activities is a letter from an outdoor clothing manufacturer who has asked the class to test different fabrics to find out which would be best to use in outdoor clothing. Indeed, it could even be presented by the head teacher who could be asking for information for a redesign of the school uniform.

Once the children have read the letter, then discussions can take place about what the term 'best' means. Here are a few suggestions:

- *Most durable*: children could test fabric by wrapping it around a wooden block and rubbing it across a rough surface such as concrete to simulate someone climbing over rocks.
- *Stain resistant*: Here fabric could be caked in mud and rubbed on grass then washed to find out which fabric still has mud and grass stains after washing. The

children could take a bowl of water and soap outside to wash the fabric, then hang it on a 'washing line' to dry, then check for stains.

- ***Waterproof***: A range of fabric could be showered using a watering-can or plastic bottle.

The whole class could go out into the school grounds to carry out all or different activities; alternatively, groups could work outside independently within a limited area. Thinking back on the strategies suggested in this chapter, groups could be asked to 'pack' their own haversacks with the equipment they need and also be given a camera to record, a stopwatch for time management, flags to indicate if and when they need help and, of course, Role Badges to encourage individual responsibility.

The Snowman's Coat

Snow is never guaranteed but if you are fortunate to have a good snowfall then take the children outside to explore. Even better would be to use Keogh and Naylor's (2000) (www.conceptcartoons.com) 'The Snowman' in *Concept Cartoon* to initiate discussion about insulation and whether or not a coat will keep the snowman warm and make it melt, make no difference whatsoever or keep the cold in and stop the snowman from melting. Prior to going out this is a good opportunity to challenge children to think about health and safety issues, from being careful on ice and not throwing snowballs to what they should wear and what they might need to take outside with them in their Science Outdoor Haversacks to make sure that they do not have to keep going back inside the classroom and making a mess of the floor!

Sarah at Castleside used the concept cartoon in the classroom and then her children built big snowmen outdoors and gave them coats, a children's winter coat, a plastic bin liner coat and no coat.

The Snowman

'The children went out and built their snowmen and tested the ideas in the concept cartoon. The children worked brilliantly in small groups, discussing issues such as making the snowmen the same size and observing them at regular intervals and also talking about how they know which snowman was melting more than the others. Photographs were taken as evidence of their investigation and when they were asked if they had learned something they all came to the consensus that they had and were able to say what it was. Results were explored and they used their experience to justify their conclusions.

To my mind this was a much more memorable experiment in building real snowmen than by being given ice cubes to test their ideas. The children were really motivated and took great delight in recounting on an hourly basis which snowman was melting the most! They also worked independently and their behaviour was excellent.

In the end I told the children that they had been doing science all afternoon and one boy piped up saying "But we've just been playing!"'

References

Futurelab (2008) *Reimagining Outdoor Learning Spaces: Primary Capital, Co-design and Educational Transformation.* A Futurelab Handbook. Bristol: Futurelab.

Keogh, B. and Naylor, S. (2000) *Concept Cartoons.* Sandbach: Millgate Publishers (available online at www.millgatehouse.co.uk).

CHAPTER 5

Health, safety and risk 'Beyond the Classroom Boundaries'

What is the goal?

All schools within the UK are required to attend to safeguarding issues to ensure that there are effective safeguarding systems and frameworks in place, and of course this relates to children working outdoors in the school grounds. Every school is unique in terms of what is 'Beyond the Classroom Boundaries' in their school grounds and it would be inappropriate to create a definitive approach to health, safety and risk. Therefore, in this chapter, we aim to offer advice concerning working 'Beyond the Classroom Boundaries' and focus on the following:

- the responsibility of adults working with children
- developing children's Personal Capabilities to enable them to work safely outdoors
- whole-school strategies for working safely outdoors.

If you and your school decide to be proactive in developing science 'Beyond the Classroom Boundaries' this should include a review of the school health and safety policy to ensure that it has a section relating to children working outdoors. Local Authorities such as the Norfolk County Council suggest that, when reviewing and possibly rewriting your outdoors policy, you should aim to include ideas from the following groups:

- the children
- all staff
- parents
- governors and managers
- the wider community (the local authority, the garden centre).

We advocate that you take time to ensure that the process is thorough and the following questions could provide a framework for review:

- Where are we now?
- Where do we want to go?

- How can we get there?
- How will we monitor success?

The goal should be to have a policy where everyone, including the children, is a partner and shares responsibility in understanding and managing the health, safety and risk inherent to learning 'Beyond the Classroom Boundaries'.

What is the reality?

While it is everyone's duty to ensure safe working practices, 'Beyond the Classroom Boundaries' requires a hierarchy of roles and responsibilities in relation to health and safety when children work in the school grounds.

Table 5.1 Roles and responsibilities in relation to health and safety 'Beyond the Classroom Boundaries'

Governors	• Ensure that documentation is in place and that it covers working outdoors, and it is monitored and reviewed regularly.
Head teacher	• Has a duty of care towards everyone working outdoors. • Implementation of the policy on a day-to-day basis linked to working outdoors. • Ensures a positive culture towards health and safety rules relating to outdoors and that they are followed by everyone.
Science leader	• Ensures that the staff are aware of relevant health and safety issues in science. • Carries out risk assessments regarding the school grounds and science equipment. • Informs staff of where science safety publications are kept. • Support staff in developing and carrying whole-school health and safety policies in science.
Staff (including teaching assistants)	• Be familiar, and comply with, health and safety arrangements for working outdoors. • Report any problems or incidents immediately. • Indicate potential risks and how they will be managed in lesson planning. • Consult science leader when in doubt. • Set a good example to others. • Develop safe practices with children and their responsibility to manage risk.
Children	Self-manage risk to make sure that they take responsibility for their own safety and the safety of others.

All schools should have a copy of the Association for Science Education (ASE) (2011) publication *Be Safe!*; preferably more than one, so that all staff have their own copy or know where to access one. Many local authorities are members of CLEAPSS and this means that primary schools in those authorities can have access to the website and extensive materials relating to health and safety in primary science, and in particular working outdoors. Both ASE and CLEAPSS welcome telephone or email queries from teachers regarding health and safety in schools. Information about both of these organizations can be found in the References at the end of this chapter.

Health and safety in primary science is generally a matter of common sense, with some exceptions relating to specific activities or equipment that are covered in the safety documentation mentioned previously. The use of school grounds should be covered in school documentation relating to safeguarding children which would include all activities where the school grounds are used, such as playtimes and outdoor PE sessions. Hazards will have been identified and action taken where necessary. The school will already have in place rules and procedures: for example, that children should not climb trees in the school grounds or children should stay away from the car park area, and this would apply for whatever reason the children are outside.

What are the options?

When children are engaged in science outdoors the risk factors change very little from when they are outside at playtime, when engaged in a PE lesson or moving around the school generally. Therefore, there are very few new health and safety issues relating to children working 'Beyond the Classroom Boundaries' in science that staff and children have to consider.

Once you have made sure that health and safety issues have been managed and actioned through the general school policy, attention can then turn towards how to support the children to work outdoors successfully and safely. There are a number of issues that should concern us in science and they are listed below:

1. Children's understanding of the concept of a hazard and of risk.
2. The use of specific equipment that might pose a health and safety issue: for example, binoculars, children should not look directly at the Sun.
3. Children collecting or handling plants and animals.
4. The development of children's Personal Capabilities in science to manage themselves and others safely when working 'Beyond the Classroom Boundaries'.

The first crucial point in developing children's understanding of safety and dovetails with their personal development of self-management skills is their understanding of the concepts of hazard and risk. There is a difference between hazard and risk, and this can be explained to children in simple terms in the context of their science. A hazard is something that can cause harm: for example, eating potentially poisonous berries from a tree, or looking directly at the Sun. A risk, on the other hand, is the chance that a hazard will actually cause harm. For example, if there were no trees with berries then there would be no risk of being poisoned; if you never look directly at the Sun then the risk of damaging your eyesight in this way does not exist. However, if you occasionally look directly at the Sun your risk increases, but if you stared at the Sun for a long time then the risk of damage to your eyes would be very high.

So, when we consider the second point, most things in life could be hazardous, and that includes using primary science equipment. However, the risks are limited if we act responsibly and use equipment safely, and increased if we misuse our equipment. When we consider the third point, plants and animals in the school grounds can pose hazards since some plants are poisonous, and climbing a tree to observe eggs in a nest can be dangerous. However, school policies should have covered these with statements such as:

- the caretaker/gardener checks regularly to ensure that no poisonous plants are growing in the school grounds
- low branches are cut so that children cannot easily climb trees in the school grounds.

Potential hazards are therefore limited and the risk of poisoning or children falling from a height is minimized. It is then up to the teacher and children, and indeed it is beneficial for children's education, to follow some basic safety rules when using equipment and working outdoors, such as:

- if you are investigating the effect of exercise on heart rate do not overdo it
- wash your hands after you have been collecting invertebrates.

Most national curricula for primary science demand that children develop their understanding of safe ways of working, adopt safe practices and manage risk in their science activities. We would expect children to progress through the primary years, with the teacher helping them to recognize risks and eventually for children to actively control risks to themselves and others.

When children work inside the classroom we expect them to work safely, so we should expect no less of them when they work beyond the classroom. The best way forward is to work with children as partners in understanding potential hazards and in minimizing risk to themselves and others.

Some schools have developed a whole-school approach to sharing responsibility with pupils. At Castleside Primary School each class created its own set of rules for working 'Beyond the Classroom Boundaries'. This was to ensure that all children were involved and that they worked at their personal ability level, as well as taking ownership of this issue. Interestingly, all the sets of rules had common statements, so by default there was parity in approach across the year groups, which gave the sets of rules a whole school feel. Once the rules had been agreed the expectation was that all children would:

- abide by the rules
- work together to make sure that whole groups kept to the rules
- take their roles seriously and would confidently challenge a classmates if their behaviour put them or others at risk.

Year 4 children and Sarah their teacher collaboratively wrote and agreed class rules for working 'Beyond the Classroom Boundaries' which are displayed indoors in every classroom and are also on view outdoors in weather proof wallets. The rules state:

- follow instructions and stay on task
- be well equipped
- walk sensibly around school
- work as a team
- contact the nearest teacher when needed

The children came up with various statements, discussed them and chose five key rules, and because they have ownership of them, they understand their meaning and are more likely to follow these self-imposed regulations.

At Shaw staff were confident in children working 'Beyond the Classroom Boundaries' if the children were in sight of an adult. However, they still considered the issue of the safety of children when they were working independently, out of sight. So they decided that, rather than allowing children to work on their own or in pairs, they would take another approach which is described below.

At Shaw, when pupils work outside, they are responsible for keeping themselves and other children safe. If they are working away from the view of an adult they usually work in groups of three, so, for example, if one gets injured, one child stays with that child and the other goes for help.

The link to Personal Capabilities, specifically teamwork, encourages children to consider and manage risks to themselves and to others in their group. It also draws on communication ensuring that children are confident to communicate concerns about safety to each other. Thus, children are engaged in self-management because they have to ensure that they take responsibility for working safely.

Developing guidelines with children prompts positive action because it allows the children greater control of their own performance.

However, if despite all the strategies something happened the children had an understanding of the protocol for getting help: for example, one child stay with the injured pupil and another goes to get help from a teacher or other adult.

An alternative approach was developed by Kay at Wheatlands Primary School and is being adopted by staff in most classes.

I worked with a colleague who was a Year 4 teacher, we developed a flag system using laminated coloured sheets of card. When the children were using the outdoor space independently they knew their parameters, that they weren't allowed to go further than X, Y and Z. If they saw the green flag in the window then knew they were free to continue their task. If they saw the amber sign come into the window they knew they had approximately five minutes to finish off. Then if the red flag went up in the window they had to come back immediately. The children were really good because every now and then I would test them and put a red one up and they would all come back, it worked really, really well. My year 4 colleague commented that it was really good how the children were mindful of what they were doing yet also kept an eye out for the flags as well. She was surprised at how quickly they responded when the red flag went up. We are both going to use it with future classes, already other staff are taking this approach on board, having seen it work with us, so I am sure that it will filter through the rest of the school.

Alternative perceptions of 'risk'

When we discuss risk in primary science, whether it is when children are working inside or outside the classroom, we usually think about health and safety issues. However, risk can relate to different aspects of out lives: physical, mental and emotional. Risk is relative to the situation and the capabilities of the individual to deal with a situation. There are many different types of risk in primary science – for example, risk can mean:

- sharing your ideas with other people because you are not sure what their reaction might be
- trying a different approach because you have never done it before
- trying something new, such as using a pooter, because you are not sure if you are using it correctly and might swallow an insect
- overcoming physical obstacles relating to walking on uneven surfaces, balance or manipulative skills
- using equipment safely, such as binoculars, and not looking directly at the Sun or making sure that you lay down on the ground by the pond when dipping for pond animals so that you cannot fall in the water
- being allowed to go outside for the first time with a science partner and to work away from the classroom without direct supervision of the teacher.

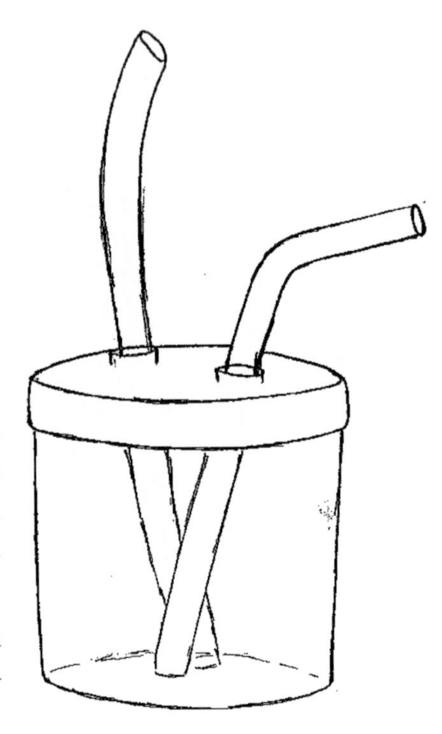

5.1 Pooter

Project schools raised an important issue about the difference between science inside the classroom and science 'Beyond the Classroom Boundaries': they were concerned with the distance it creates between pupils and teacher. However, we can view the distance that the school grounds creates between teacher and children positively rather than as a worrying situation. That small distance can help to develop children's independence and management of personal risk in different ways: emotional, physical and mental. In the classroom the teacher might be oversupportive because it only takes a few steps to get to a child. Whereas when working in the school grounds it may take a little longer to reach children, which means that the teacher is more likely to wait and observe, decide who to go to and why. The children realize that attention is unlikely to be immediate and are more likely to take responsibility and to solve the problem themselves.

In the example below the concept of risk when working 'Beyond the Classroom Boundaries' takes on a different dimension and serves to remind us of the statement made at the beginning of this chapter, that 'every school is unique'.

Woolley Wood is a special school where the staff worked to overcome a range of issues relating to children engaged in science 'Beyond the Classroom Boundaries.'

One key issue for the staff was that not all classes had immediate access to the outdoors and many of the children are non-ambulant or have limited independence when walking. So, staff had to think carefully about the challenge of using the outdoors more frequently.

The idea of risk is relative to the individual school and child, and when working outdoors risk might not be directly related to science itself, but to the emotional or physical abilities of the child.

During a science session outdoors, one of the children decided to walk up a hillock to find some worms. Suddenly staff heard her calling because she was unsure of how to get down from the small hillock in the school grassed area. In the confines of the classroom the usual response was for the child to wait to be helped and staff to take the few paces towards the child to assist her.

However, working outdoors meant that there was often a greater distance separating the child and adult, and consequently staff could not get to the child immediately.

In the time it took the staff to reach the child she had thought about the alternatives and then to her delight, and that of staff, negotiated her way down the hillock.

For this special needs child this was a personal milestone, and she was very proud of her own achievement. She found herself in a challenging situation – it was in fact a moment of risk, but also one of real learning which she would not have faced in the classroom. Interestingly, she then placed herself back in the same position and repeated getting down from the hillock a number of times, showing others (and herself) that she could do it and celebrating her newly found independence and confidence. The hillock was no longer a hazard and staff no longer saw her as being at risk. Risk (and, let us be clear here, a calculated one) can be used positively as a means of providing opportunity for developing Personal Capabilities.

What will you do?

If we go back to the beginning of this chapter then the answer to 'What will you do?' must be to ask and answer the following set of questions.

- Where are we now?
- Where do we want to be?
- How can we get there?
- How will we monitor success?

The most successful approaches have been where staff and children across the school have been involved. The next section suggests some practical ways to involve children in using and developing their Personal Capabilities to work independently and safely 'Beyond the Classroom Boundaries'.

Practical ideas for developing children's ability to work safely and independently 'Beyond the Classroom Boundaries'

Understanding hazards and risks

As you begin to introduce children and indeed staff to working 'Beyond the Classroom', the first issue is an understanding of the concepts of what is a hazard and what is a risk.

Think–pair–share
Ask children to discuss with their science talk partner the meaning of the words 'hazard' and 'risk'. Encourage them to think about where they have heard the words used, and to provide examples from their own lives. They could draft definitions on their individual white boards, which could be changed at a later date if they think their understanding has changed.

Make a foursome
When children are happy with their definitions then ask them to join with another pair, to share their ideas and then to use their definitions to create an agreed whole-group definition. This will certainly draw upon a range of Personal Capabilities such as listening to each other, turn taking, compromising and respecting ideas from others in their group.

Share with the class
This is the point where groups share their definitions with the rest of the class – and it is here that the teacher can help the class to recognize similarities in their ideas and help children to come towards a class concensus on the ideas of hazard and risk.

Children carrying out a risk assessment of the outdoors area

Once children understand the idea of 'hazard' and 'risk', why not take them outside to make a risk assessment of the school grounds. Tell them that they are going to a 'Hazard Spot', where they have to decide what in the grounds could be a hazard, their reason/s and how they can minimize the risk of anyone getting hurt.

You could also challenge children to think about how they would record the hazards that they spot. Children might, for example, create a table of their observations, use a camera to photograph potential hazards, or use an Easispeak microphone.

5.2 Hazard spot

Figure 5.1 Children's Risk Assessment Card

OUR SCIENCE RISK ASSESSMENT CARD

BE SAFE!

Our activity

____

____

____

____

HAZARD	Why is it a hazard?	How do we keep ourselves safe?

Signed________________________.

Date ________________________

Risk Assessment Cards/Forms

One of the key aims of this book is to encourage children to become independent learners who take charge of their science. Thus, when children work 'Beyond the Classroom Boundaries' they know that they need to stop and think about safe working practices. Then if they think that what they are going to do outdoors might pose some risk, the children make the decision to complete a Risk Assessment Card, and then hand it to the teacher or adult they are working with to 'sign off'.

Of course, such a system needs to be scaffolded where the teacher helps to set up this way of working and with adult support in the initial stages, children learn to use the cards. Gradually as their personal understanding in science develops, and their appreciation of hazards and risk increases they become more capable of making their own decisions.

RISK ASSESSMENT CARD

NAME/S:

CLASS:

RISK	WHAT TO DO

5.3 Risk Assessment Card

Health and safety monitors

In an earlier section of this chapter we indicated that it is the school governors and the head teacher who have ultimate responsibility for safeguarding issues around the school. We also suggested that on a daily basis everyone should contribute to health and safety, so it is appropriate that a school might think about appointing 'Health and Safety Monitors' who regularly walk around the school grounds and spot potential hazards. This is an excellent way to use and develop Personal Capabilities such as, teamwork, responsibility, communication and motivation. For example, children could report a broken fence, glass or litter thrown over into the school grounds, or loose stones that someone might trip over. The emphasis is on children identifying and reporting risk and hazard, which the school then attends to. In appointing 'Health and Safety Monitors' it is appropriate to give them badges so that everyone knows that they can report perceived risks; recognizable badges also promote the children's self-esteem.

Hazard warning signs

This could be a whole-school or year-group project where 'Hazard Warning Signs' are designed, made and positioned strategically around the school grounds. Children could be given the responsibility to locate hazards and then design hazardous warning signs, and possibly make them or prepare a design brief for their manufacture by a specialist, who could work with children to construct them.

'Safety in Science Outdoors' poster

Why not have a school competition where the children create a 'Safety in Science Outdoors' poster and the wining poster is printed as a weatherproof notice to promote safe working in science outdoors.

Working Safely 'Beyond the Classroom Boundaries' leaflet

Older children could create a leaflet that explains to children new to the school or to parents, school governors and so on the difference between a hazard and a risk and the many ways that children work safely when they are outdoors in the school grounds. Their leaflet could include:

- spot the hazards
- word searches
- 'what should you do?' sections
- crosswords
- rules for working outdoors
- match the photograph to the hazard.

References

ASE (2011) *Be Safe* (3rd edn.). Hatfield: ASE.

Norfolk County Council Early Years Outdoor Learning. Available online at www.schools.norfolk.gov.uk/myportal/custom/files_uploaded/uploaded_resources/5381/earlyyearstoolkitfinalversion.pdf

Contacts

Association for Science Education
College Lane
Hatfield AL10 9AA

Tel: +44 (0)1707 283000 email: info@ase.org.uk

CLEAPSS®
The Gardiner Building
Brunel Science Park
Kingston Lane
Uxbridge UB8 3PQ

Tel: +44 (0)1895 251496 email: science@cleapss.org.uk

CHAPTER

6

Progression and a review of your own progress

What is the goal?

Now that you have been introduced to the ideas related to working in science 'Beyond the Classroom Boundaries', it seems appropriate to suggest a pause to think and to consider the steps that could be taken towards competency and confidence in using the outdoors. In this chapter we introduce those progressive steps, known as the 'Escalator' (see Figures 6.1 and 6.2) Of course, movement does not happen overnight; in the Escalator we present some thoughts around what it could potentially 'look like' and 'feel like' for teachers and learners, to move from practice which does not exploit working outdoors in science, to one where it is a regular and routine part of classroom life.

This chapter is designed to stimulate self-reflection and reflection on children's abilities and skills. You could use the Escalator on your own or with other teachers to review your current position and aspirations and to stimulate discussion with colleagues, governors, parents and indeed the children if you like, and as importantly noting where you are and where you want to be; to objectively consider what it will take to move progressively up the Escalator. Throughout the chapter you will be referenced back to key sections of the book where approaches and resources are explained more fully.

What is the reality?

What works well for one person will invariably not be quite the same for the next. Progress is determined by where you start and is judged by where you hoped you would get to. In order to gauge where you are in your thinking and practice at the moment, we offer you two types of 'Escalator' – an 'Escalator of children's progress in using the outdoors for learning (Fig. 6.1)' and an 'Escalator of outdoor learning entitlement for teachers in schools (Fig. 6.2)'. The Escalators offer a six-stage model from what could be teachers or pupils who currently do not use the outdoors to others who use it as an integral part of teaching and learning and those who have fully grasped and exploited its potential. Each stage is illustrated by a range of descriptions that aim to support reflection

Figure 6.1 Escalator of children's progress in using the outdoors for learning

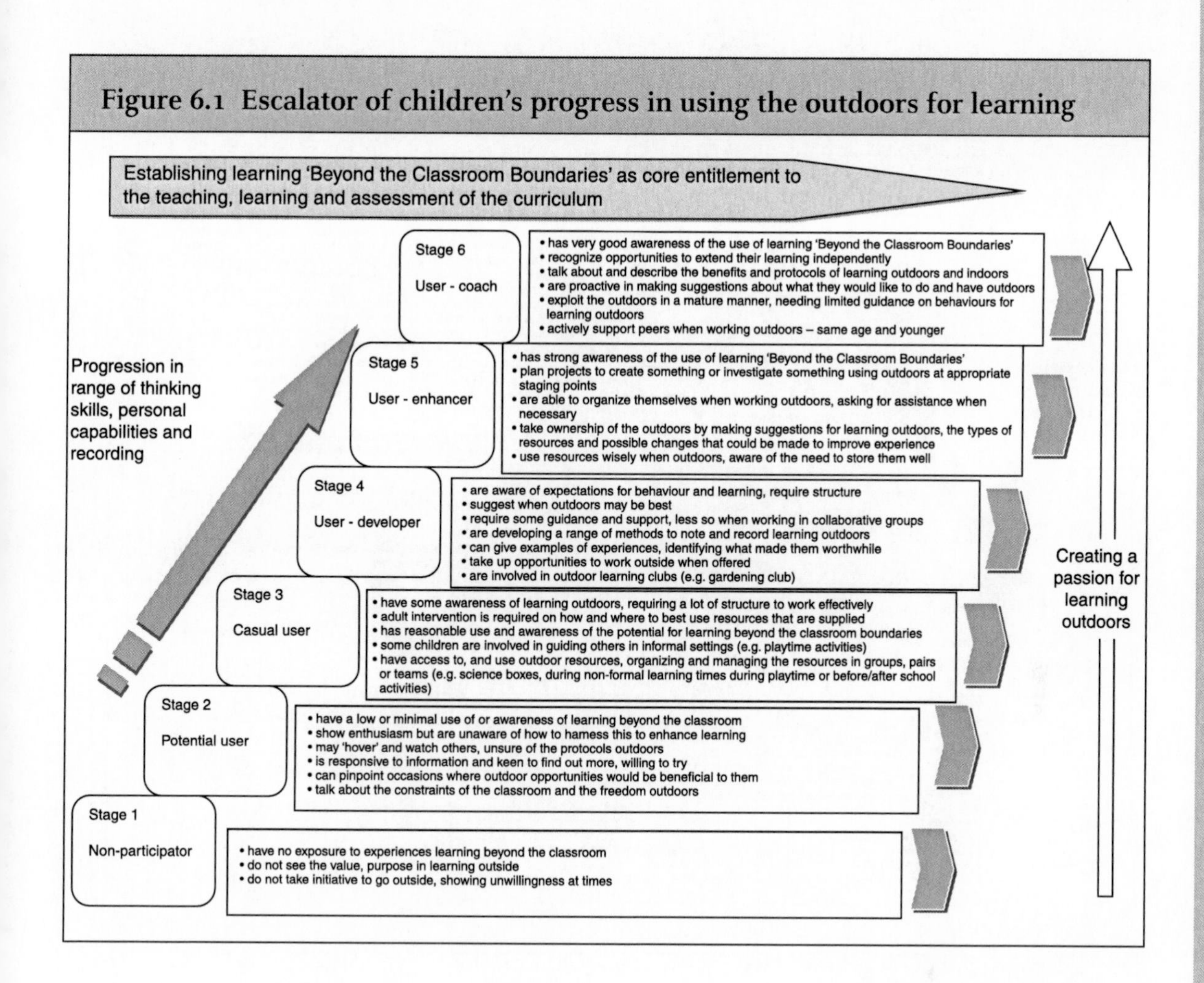

and assist us in identifying not only where we are but what we do particularly well at and what we might aspire to do more of.

So in order to consider your 'reality', take some time to look at each of the Escalators. The first considers your children and their competency at working outdoors and the second your position as a teacher and the provision you offer. Try to be as honest in your appraisal of yourself and in appraising your pupils. Do not make excuses or look for reasons to justify your position, just make as honest a judgement as you can using a 'best fit' approach. You then have a great starting point to consolidate and improve teaching and learning in science 'Beyond the Classroom Boundaries'.

What are the options?

There are a large range of options open to us and this book could be viewed as the tip of the iceberg in terms of where the developments in outdoor learning could lead. The options open to you will depend on where you and the children are on the Escalator, the children's age and ability, school resources, grounds and support, and so on. An attempt has been made in the table below to give some pointers as to what could assist the progression from one stage to another within the Escalator model. You will find references to

Figure 6.2 Escalator of outdoor learning entitlement for teachers in schools

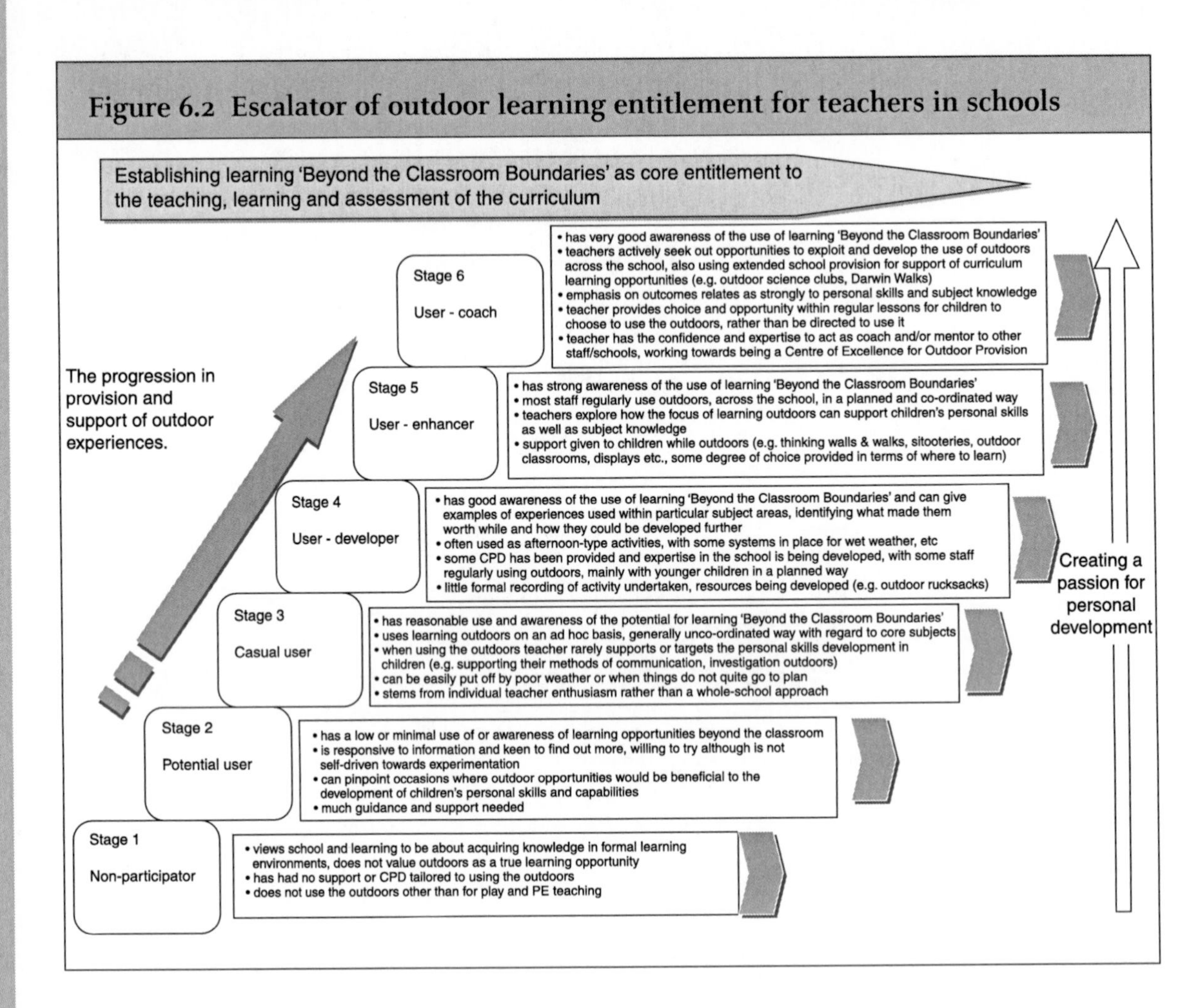

other pages/sections in this book where activities, resources or experiences are explained in more detail.

Where to start?

Before moving blindly into the usual routines when starting an initiative, such as asking yourself, 'What can we do?', 'When shall we do it?', 'How much will it cost?', 'Who can help?', take a little time to think about what you are actually trying to achieve by expanding teaching and learning into work 'Beyond the Classroom Boundaries'.

Take the GROW approach that you have come across in each of the chapters in this book. Consider:

- What is my/our goal?
- What do you want to gain from this for the children's personal development? What do you want to gain for the teaching of science? What do you want to gain for yourself as a professional?

Sarah's Goal – All weather science

Sarah at Castleside regarded her main goal as making sure that all teachers used the school grounds in science across the year, not just when the weather was fine. She aims to develop activities outdoors to encourage staff to use the grounds for lessons and children to explore in science at playtimes.

Carol's Goal – Seamless learning

At Shaw Primary School, Carol realized that the school grounds, which were extensive and excellent for science outdoors, were not used to their fullest potential. Her goal was for there to be no boundaries between the classroom and the outdoors in science. Her vision was the seamless movement of children who as partners with the teacher would be engaged in decision making about where they thought their learning in science should take place.

Once you have identified your goals, and used the Escalators to consider your 'reality', you will be in a much stronger and more focused position to consider the options.

Now that you are at the stage to consider making changes, the key thing is to be open minded and not to limit yourself or your children. Create a table similar to that in Table 6.1, for yourself, leave your options open, then take some time to talk them over with colleagues. Do not 'just settle' for what may be easiest, but set aspirations that challenge, push practice forward and even takes a few risks with your teaching and the children's learning.

Here are some suggestions:

Table 6.1 Moving forward

From escalator stage	What teachers could do and pupils could be involved in ...	Ideas/resources that could support
1–2	Identify the opportunities that are currently provided in the school grounds and the facilities/resources already available. Make a list of areas and locations that have the potential of being improved or better tailored to outdoor learning. Consider the practicalities of working outdoors – what health and safety issues arise, what will the children need to be equipped with to work outside effectively, etc.?	See Chapter 2. See Chapter 4 to review the issues around managing the outdoor spaces and Chapter 5 to help address some of the health and safety safeguarding issues.

(Continued overleaf)

Table 6.1 Continued

From escalator stage	What teachers could do and pupils could be involved in . . .	Ideas/resources that could support
	Talk with the children about where they like to learn and why. Encourage the use of outdoor science equipment by putting out resources at informal learning times, e.g. playtime, before/after school clubs. Give responsibility to children for looking after the equipment, showing other children how to use it, organizing it, etc.	See page 86 to identify ideas such as Science Boxes.
3–4	Review your current science planning for your year group or across the school. Does the current topic/theme plan link in with possible outdoor occurrences – the seasons, daylight hours, growth of plants, diversity of living things, etc.	See chapter endings to see examples of how to use the school grounds in science.
	Explore what CPD opportunities are available to yourself and your staff. Consider who may be best at leading this area of development with you – maybe a colleague from another Key Stage, a teaching assistant, group of parents, a local scientist, etc.	See Chapter 5 to find a list of useful websites and groups that could provide inspiration and ideas to support your goal.
	Consider what visual and verbal protocols would support children learning outdoors, with or without adult supervision.	See Chapter 4 and think about using bibs, cones, walkie-talkies, etc.
	Observe children working in pairs, small groups and teams during structured activities outdoors or even in the school hall. Consider which of the areas of Personal Capabilities they are strong in and which you feel they need support with. Talk with them using simple AfL strategies, such as thumbs-up-sideways-down about how they feel when working in a different type of space and what type of things would help them learn better.	See Chapter 3 to read more about the Personal Capabilities children use when working in this way, and how simple structures can support their behaviour and learning.

From escalator stage	What teachers could do and pupils could be involved in . . .	Ideas/resources that could support
5–6	Discuss with colleagues how the range of outdoor learning opportunities can be refined to enhance the development of children's Personal Capabilities. Consider when it would be appropriate for children to be given greater autonomy and choice in deciding for themselves when to work within the classroom and beyond it.	See Chapters 3, 4 and 6. Specifically consider if any of the teacher ideas could be of inspiration to you.
	If aspiring to be a Centre of Excellence, how can the whole school embrace this style of learning. What whole-school developments could take place beyond science lessons (e.g. the Darwin Thinking Path, Thinking Posts, etc.)	See Chapter 9 Darwin's Thinking Walk.
	Consider which groups of children could be 'Beyond the Classroom Boundaries' visitor guides, or facilitate a group of children who consider, decide on and carry out a project to improve provision in the school grounds for science.	See the end of this chapter for ideas.

What will you do?

So maybe we have arrived at the million-dollar question. . . . What are you actually going to do?'

You are clear on your *Goal*, you have reviewed where you think your school, teachers and children are, your *Reality*, and you have read about and decided on the range of *Options* open to you. So what next?

Now you are at the stage of deciding the '*What*' as in 'What will we do?' Do you make the final decisions on your own, with a colleague or even with the pupils? However you decide to do it, the key thing is to be ambitious and to challenge yourself while having a clear sense of how it can be achieved, in what timescale, who would be involved, and how you will measure success. After all, you will want to share your success and be able to identify what worked well and the real gems of experience that help to tell the story. Indeed, to understand why things did and did not work is essential for you as a professional learner and curriculum developer.

You may consider some of the following as ways to catalogue what has been happening and its effects:

- *A blog*: for you, other teachers, the children and perhaps parents.

- *A working wall*: an interactive, collaborative display where all groups in the school can display photographs, pieces of work, lesson ideas, reflections, just think if this could even be outdoors!
- *A big book*: class-based or school-based, using pictures, diagrams, pieces of writing, investigation write ups and so on.
- *Outdoor working party*: group of children (similar to the School Council) who advise on outdoor learning activities, review experiences, take the lead in organizing outdoor equipment, provide a forum for reflection and undertake pupil surveys. Children could produce a 'Newsletter' communicating progress to other children, teachers, school governors and parents.
- *Assemblies and reward certificates*: focusing on 'What Progress We've Made' learning outdoors.
- *The school website*: with photographs, news items and informing parents of forthcoming activities and events 'Beyond the Classroom Boundaries'.

Audit Map of Shaw Primary School Grounds

Carol at Shaw began with a huge audit map (see Chapter 2) that took up a whole display board in one of the corridors. This was used to show the progress of the development and use of the outdoors in science. It served to celebrate new science areas in the school grounds and how they were being used and to let everyone know what would be developed next. This was an excellent way of communicating to a range of people from the children themselves to staff, parents, school governors and any other visitors to the school.

The following (Tables 6.2–6.6) are examples of frameworks that you could use to guide your planning, within one class or across the school.

Table 6.2 A whole-school half-termly planner indicating science topics and whole-school developments 'Beyond the Classroom Boundaries'

	Foundation/ Reception	Y1	Y2	Y3	Y4	Y5	Y6	Whole School
Autumn 1								
Autumn 2								
Spring 1								
Spring 2								
Summer 1								
Summer 2								

Table 6.3 A year group monthly planner indicating general provision across the curriculum 'Beyond the Classroom Boundaries'

Month	Learning experiences	General resources (including people)
September	Science – Literacy – Numeracy – Etc.	
October		
November		
December		
January		
. . .		

Table 6.4 A planner for teachers to use indicating short-term and longer-term goals

	Quick Win 1	Quick Win 2	Longer-term goal
What is your goal?			
What will you do?			
What will you need?			
How will you evidence this activity?			
How much effort will this require? (1–5 scale)			

Table 6.5 Using the GROW approach to map the experience

Goal What is your goal?	
Reality What is your reality?	
Options What are your options?	
Will What will you do?	

Table 6.6 A written reflection grid for teachers

1 Describe the lesson/activity that took place 'Beyond the Classroom Boundaries'. 2 How did you feel about the process and outcome? 3 What was good and what needed adjusting about the lesson/activity? 4 What have you learnt? What do you take away from this lesson/activity? 5 If you did it again what would you do?

The real power behind any of the suggested frameworks, or indeed the use of the Escalator is, of course, the level of honesty and objectivity of the user. In many cases it will be necessary to broaden one's understanding of what is actually happening in other classrooms in the school. Take this opportunity at this point in your reading of this book to stop and begin to objectively judge the stage of your own work and that of other teachers, and to be frank about the true impact this is having on pupils. Talk to teachers, talk to pupils; it is through such talking processes that you as an innovator will really glean a sense of the status quo in your school.

Practical ideas for teaching plants 'Beyond the Classroom Boundaries'

'Beyond the Classroom Boundaries' offers a range of opportunities to develop children's understanding of plants as well as the prospect of helping to progress children's Personal Capabilities. In this section of the chapter we have sought to provide activity suggestions for children engaged at each stage of the Escalator to help them and indeed to support staff in moving children forward.

Stage 1: Non-participator

Children have no exposure to experiences learning beyond the classroom; they do not see the value, purpose in learning outside; they do not take initiative to go outside, showing unwillingness at times; they adopt the view that learning is about acquiring knowledge in formal settings, and that the outdoors is just for play.

For these children early steps to encourage them to be outdoors are key. Try short, structured tasks: for example, in pairs to go and photograph as many plants as they can find in 10 minutes. They can then be asked to select their two favourite pictures and to create a display by encouraging them to mount the photographs outdoors, perhaps on a fence, using plastic wallets and placing keywords related to the plants inside or around the wallet using waterproof markers.

This is also a good point to introduce the 'Science Outdoor Boxes' to children. Take the class outside and show them where the boxes are and take out all the equipment and encourage children in groups to think about what the equipment can be used for and what they might do with it during play and lunchtimes. Also ask children to think about the rules that were explained to the whole school in assembly when the boxes were first introduced to the school.

Over the first couple of weeks do have 'share times' for a couple of minutes after playtimes so that children can talk about what they have used and what they did to celebrate and value children using the boxes.

In pairs, supported by a teacher, assistant or parent helper children could explore the school grounds with the aid of an identification key. They could take part in a plant hunt where they have to spot as many different varieties of plants as they can. Children could then compare results between different group 'hunts' and collate information to be used to draw conclusions about plants in the school grounds.

Stage 2: Potential User

Children have a low or minimal use of or awareness of learning beyond the classroom; they show enthusiasm but are unaware of how to harness this to enhance learning; they may 'hover' and watch others, unsure of the protocols outdoors; they are responsive to information and keen to find out more, willing to try; they pinpoint occasions where outdoor opportunities would be beneficial to them; they talk about the constraints of the classroom and the freedom outdoors.

When beginning science activities discuss with the children whether they think this activity is one that would be better to do outside rather than inside. For example, in plant topics we often grow beans to answer questions such as:

- Which plant grows the tallest?
- Where is the best place for the plant to grow?
- How does the amount of light or water affect the plant?
- Will the plant grow better if we feed it?
- What will we use to feed the plant? For example, different kinds of plant food, tea leaves.

Then ask the children to think about how they would organize themselves to carry out this activity outside: for example:

- What will they need? (Equipment could be placed in their science haversacks.)
- Where will they plant their seeds/plants?
- How will they label their seeds so that they know which seeds/plants are part of which test?

- Will it be all right just to use one seed/plant for their test? What happens if it dies?
- How will they record their results? How can they share their observations with the rest of the school outside in the school grounds?

This is a great way to begin by presenting children with a more structured activity, which is mostly outdoors. It also means that as long as the area is accessible at play- and dinnertimes, children might be interested enough to visit the plants and see how they are progressing.

Stage 3: Casual User

Children have some awareness of learning outdoors, requiring a lot of structure to work effectively; adult intervention is required on how and where to best use resources that are supplied; have reasonable use and awareness of the potential for learning beyond the classroom boundaries; some children are involved in guiding others in informal settings – for example, playtime activities; have access to, and use, outdoor resources, organizing and managing the resources in groups, pairs or teams – for example, science boxes, during non-formal learning times during playtime or before/ after school activities.

Here the children should be comfortable with using the science boxes outdoors for their own explorations and enjoyment. The aim here is to introduce children to a range of activities that they can then repeat on their own during play- and lunchtimes. In this activity we suggest carrying out a 'Smart Hunt' modelled on the materials from Smart Science (see www.personalcapabilities.co.uk) where children:

- work in teams of 3–6
- delegate roles within the team (e.g. recorder, timekeeper, photographer)
- think about health and safety issues (e.g. look after each other, do not eat plants).

The teams are directed to different areas in the school grounds to explore with their senses the variety of plants that they can find. This activity encourages increasingly independent work, in a semi-structured way, assisted by group roles and suggested response sheets and timescales. Building autonomy at the 'casual user' stage will prompt ownership and personal responsibility for learning outdoors. The children in this activity will be asked to:

- work in the area given
- use equipment such as, cameras, hand lenses, rulers and tape measures, identification sheets or books

- locate and identify the plants in the area that they have been given
- log the different plants in that area using a digital or video camera
- make notes on where the plant is growing, height and spread of plant, any animals living on or around the plant, and what it feels and smells like.

Back in the classroom the groups of six can then work in pairs to research information about the different plants which can then become, for example, part of a big book, or a set of fact files that might be placed outdoors. Children should be asked what they need to find out about the plants such as life cycle, medicinal use, or any stories behind plants.

Let the children know that if they want to use the hand lenses and identification books or charts during play- and dinnertimes they can do so and if they find any new plants to add them to the fact file or big book.

Stage 4: Developer

Children are aware of expectations for behaviour and learning, and require structure; suggest when outdoors may be best; they require some guidance and support, less so when working in collaborative groups; are developing a range of methods to note and record learning outdoors; they give examples of experiences, identifying what made them worthwhile; they take up opportunities to work outside when offered; they are involved in outdoor learning clubs – for example, gardening club.

Encourage pairs to compile a list of questions that interest them about plants; for example:

- Which are the tallest plants in our school?
- Where are the most plants found?
- How many different shades of green can be found among the plants in our school grounds?

By encouraging such questions using question stems (see Chapter 1) pairs of children could think about their questions and select one to explore in depth and over time – for example:

- How does the plant change over time?
- What is the pattern of its growth?
- How does the plant fit into the food chain, web?
- What kind of invertebrates live on or around it? Are they useful or destructive?
- How does the plant reproduce? How can we reproduce this plant? Collect seeds, take cuttings and so on.

If the school has a vegetable garden area then children could become 'Young Horticulturalists'. Give them a badge with this title or get them to make their own, they will really love wearing it and explaining to others what it means.

Ask children to research what a horticulturalist is and does. They should find out that horticulture is all about plant science, how to grow plants and cultivate them. The school garden provides a great opportunity for children to extend their understanding of plants and to develop an increasing independence to identify their own questions, set up their investigation and follow it through collecting evidence. At this stage they should be able to recognize the need for a good number of trial plants to be placed in any one condition: for example, due to other issues that affect the plants. Very often this type of activity is undertaken indoors yet, of course, the more natural outdoors setting (although more challenging to control) is more authentic. So children might have seed and plant trials which could include:

- planting one type of seed but from different seed producers to find out which seed producer is best by working out the percentage of seed germination (easy to do if they plant 100 seeds, then the number that germinate, say 75, becomes, in this case, a 75 per cent germination rate)
- finding out if depth of planting affects plant growth
- investigating whether the use of a cloche speeds up germination and plant growth.

The findings from these investigations can then be shared with other groups using the garden to grow plants which means that they have to communicate findings, perhaps to the school gardening club but also that their work is put to an authentic use. It is also more likely that children will spend some of the play or dinner times checking and recording the progress of their plant trials.

Stage 5: User Enhancer

Children have strong awareness of the use of learning beyond the classroom boundaries; they plan projects to create something or investigate something using outdoors at appropriate staging points; they are able to organize themselves when working outdoors, asking for assistance when necessary; they take ownership of the outdoors by making suggestions for learning outdoors, the types of resources and possible changes that could be made to improve experiences; they use resources wisely when outdoors, aware of the need to store them well.

Some children might be encouraged to take on the mantle of expert as Ecoconsultants. They could consider what changes or improvements could be made to the school grounds and manage a small project – for example:

- Identifying the biodiversity in the school grounds and finding ways to improve it, for example, few if any butterflies are seen in the school grounds so which plants should be introduced to encourage this invertebrate?
- Developing a sensory garden/area where children are able to touch, smell and hear plants as well as see them.
- As part of their projects they will have to create a plan which should include any costings if new plants are required and also be able to present their ideas to an audience – for example, School Council, at assembly or school governors.

Stage 6: Coach

Children have very good awareness of the use of learning 'Beyond the Classroom Boundaries'; they recognize opportunities to extend their learning independently; they talk about and describe the benefits and protocols of learning outdoors and indoors; they are proactive in making suggestions about what they would like to do and have outdoors; they exploit the outdoors in a mature manner, needing limited guidance on behaviours for learning outdoors; they actively support peers when working outdoors, whether they are the same age or younger.

There will be children in the school who could be designated as 'School Grounds Ambassadors' and be asked to take visitors around the outdoors area to explain how they used it in science, different areas, plants and so on. In terms of Personal Capabilities this is an excellent approach and many of the project schools are working towards developing pupils so that they are confident and able to be placed on a rota of 'School Grounds Ambassadors'.

Another idea which is adapted from the early years version of this book is to develop outdoor story sacks. These are a simple, yet versatile, resource for children which contain a story that has a link to working outdoors in science and equipment to encourage children to explore some of the elements from the story in the sack.

The children who are at the 'Coach' Stage would be able to take a small group of younger children outside and tell the story from the story sack and ask questions and prompt discussion with their audience and then guide the group to explore and carry out the activities suggested in the story sack. Some suggestions for books and resources for such story sacks are listed in Table 6.7.

Table 6.7 Outdoor science story sacks

Book	*Resources and activity ideas*
The Snail and the Whale by Julia Donaldson	• Snail identification sheets/cards, magnifiers. • Children find snails around the school grounds, use magnifiers, count snails, sketch and photograph snails, record observations on the Outdoor Whiteboard 'Great Snail Hunt'.
Quiet! by Paul Bright	• Boom whackers, string telephones, clucking cups, card megaphones, wooden and metal spoons. • Children explore tapping different things around the school grounds to find out what sounds they make.
Jasper's Beanstalk by Nick Butterworth	• Plant pots, bags of compost, plastic trowels, bean seeds. • Children use the equipment to plant and care for seeds outdoors.
Stick Man by Julia Donaldson	• Objects made from wood, rulers, pencils, spoons, toys, a camera. • Children carry out a wood hunt around the school to find out as many things as they can that are made from wood, take photographs of them for their own *Stick Man* book.

CHAPTER 7

Resourcing science 'Beyond the Classroom Boundaries'

What is the goal?

In this chapter we will consider how to resource science 'Beyond the Classroom Boundaries'. It is useful to remind ourselves what we are trying to achieve, so we will return here to the basic philosophy behind science outdoors, this will help us to develop appropriate resources and associated practices. We are aiming to:

- use the whole-school outdoor environment to access the primary science curriculum
- develop children's Personal Capabilities.

A carefully selected range of resources for outdoors can help to motivate children to explore activities either on their own or with other learners. In choosing resources the teacher should aim, as much as is feasible, to provide items that do not require an adult to act as mediator. In this way children are encouraged to work independently outdoors, making their own decisions about whom, how and what they work with. As children become more experienced and confident in using the resources they will become self-regulating and increasingly motivated to return again and again to explore the outdoors when they see it is appropriate and of benefit to their learning. This is of course a highly ambitious target, but one to which we should all aspire.

Do not forget that children enjoy sharing their experiences, ideas and products, so strategies to support communicating and developing a positive self-image in working outdoors are vitally important in enriching the development of science knowledge and understanding. For example, it is useful to ask questions such as:

- Which resources will be supportive of learning and provide novelty and added stimulus to learning?
- Which resources will challenge the children to show their learning in creative ways?

After all, being outdoors should be fun, something special, as well as educationally challenging.

What is the reality?

In many of the project schools teachers and pupils have described how the outdoors is not used to its full potential as a resource in its own right (this is particularly so in the upper primary phase). Few schools dedicate time or professional development to discuss and develop specific resources for supporting outdoor learning, let alone science 'Beyond the Classroom Boundaries'.

When science leaders were asked about the issues relating to resourcing science 'Beyond the Classroom Boundaries', the following were typical of responses to this question.

It is not a priority in school and so there is no money for it.

I don't have the time to put together and look after the resources.

It will be difficult to organize and get everyone to look after resources outside.

We can't leave the resources outside it isn't safe.

What are the options?

What resources can we put outdoors to sustain independent activity?

While moving a school towards working 'Beyond the Classroom Boundaries' in science will require some funding, sometimes it is simply the reorganization of existing resources that can lead teachers and children to dipping their toe in the water. In this chapter we share the creative responses from teachers in project schools who have developed ways to provide teachers and children with access to materials and equipment to support and sustain primary science outdoors.

For many schools the first step in resourcing science 'Beyond the Classroom Boundaries' was to develop an 'Outdoors Science Kit' or 'Outdoors Science Box' that provided support for children to manage their own learning, in lessons and during playtimes. Here we are engaging children in the Personal Capability of 'self-management', and to do this we should be looking towards establishing resources that will encourage children to:

- explore their scientific understanding of their immediate environment
- make their own choices about what they want to do and use

- use a range of scientific equipment safely
- record and communicate their experiences in a range of ways should they wish to do so
- organize, plan and take responsibility for how they will carry out a task
- think about how they learn as well as what they learn.

Storing resources outdoors

In all the schools involved with this project a common issue was, 'How can we store equipment and materials outdoors?' The idea of an 'Outdoors Science Box' originally came from Sue Harrison at Cheveley Park Primary in Durham. Her simple, but very effective, solution has been taken on board and developed further by other schools.

Outdoors Science Boxes

Sue's idea was to have one or more large plastic trolley boxes that could be easily pushed by pupil 'Science Monitors' outdoors, and remain there until the monitors brought them back in at the end of the day.

In using this approach Sue considered health and safety issues and made sure that children were not lifting or carrying heavy boxes and that the contents were safe items and kept dry by storing them in a weatherproof box.

Sue used a monitor system so that children were responsible for the outdoor resources; she also gave them the role of checking the contents and retrieving any items not returned to the box. This also meant that staff did not have the added burden of managing the resources on a daily basis.

The children were trained in looking after the resources and informing Sue when items had been lost, damaged or needed replenishing. A rota ensured that different children across the school were given the opportunity to be in charge of these resources and to develop associated Personal Capabilities.

Personal Capability self-management: to be responsible for the things around me.

Sue's idea provides opportunities for children to feel that they are in charge of what they do by taking responsibility for what they need, how it is maintained and managed. This necessary life skill is what many parents often strive for at home and in-school experience can only complement such endeavours.

What could an 'Outdoors Science Box' contain?

The science equipment and material should reflect the potential of the grounds and the needs of the children. Storage should be clear, attractive and imaginative to encourage children to use the kit.

Shaw Primary School Science Boxes

Carol at Shaw Primary School developed a number of Outdoors Science Boxes. Here is an example of the contents of one of them. These boxes were initially developed to support children exploring the school grounds outside of formal science lessons. You will notice that the contents have been photographed and a laminated checklist has been created so that the 'Box Monitors' can make sure all the contents have been returned to the box. Carol involved the children in this process which was an ideal way to encourage ownership of the boxes from the start.

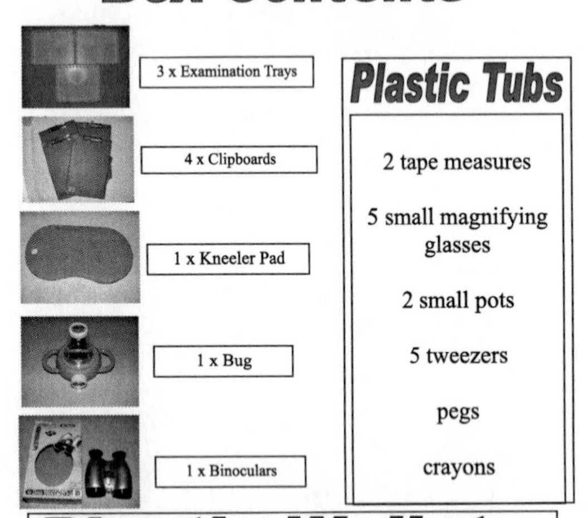

Box Contents

3 x Examination Trays

4 x Clipboards

1 x Kneeler Pad

1 x Bug

1 x Binoculars

Plastic Tubs

2 tape measures

5 small magnifying glasses

2 small pots

5 tweezers

pegs

crayons

Plastic Wallets

1 x coloured paper
1 x white paper
1 x activity cards (blue sticker on the back)
1 x Identifying Book and poster
1 x pencils, sharpener and eraser
1 x 2 Red Magnifying Glasses and Fresnel Mirrors
1 x colouring pencils
2 x bug pots
2 x mirrors

7.1 Laminated Outdoors Science Box checklist

In this example Sarah decided not to have an Outdoors Science Box but Outdoors Science Haversacks. She thought that these would:

- motivate children to want to work outside
- encourage children to take responsibility of resources
- encourage individuals and small groups of children to work independently
- support children in planning and taking responsibility for how they will carry out their tasks
- encourage children to think about how they learn as well as what they learn.

The 'Outdoors Science Haversack' concept offered a flexible approach for staff, who, if they wanted to, could change the contents of the haversack for different science activities outside.

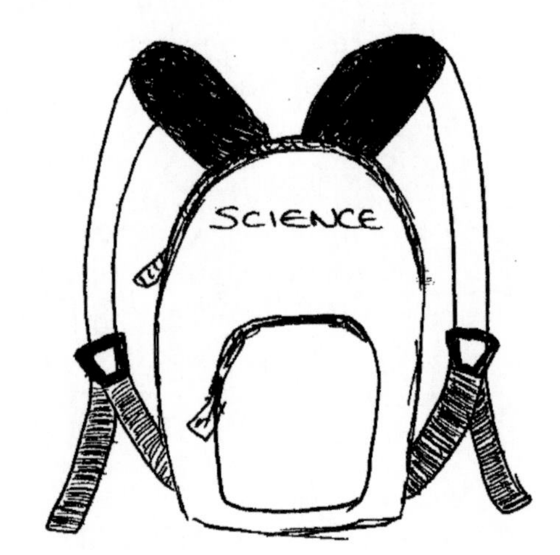

7.2 Outdoors Science Haversack

Sarah's Science Haversacks

Sarah decided to develop the idea of an Outdoors Science Haversack that children would take with them whenever they worked outside. One person in each group would take the haversack out with them and be in charge of checking the contents prior to going out, and then before returning to the classroom to ensure that everything on the laminated checklist was present.

In her orginal haversack Sarah included:

- chalk
- stopwatch
- digital camera
- Easispeak microphone
- magnifiers
- clipboard
- paper/notebook
- pencil
- sharpener
- pooters/collecting pots
- identification keys for invertebrates and plants.

Help Flags

Teachers from Sheffield schools discussed the idea of numbering each haversack and putting a set of three flags, green, amber and red flag, inside each one. The flags were designed to be used by the children to show the teacher if they need support:

- green – we are doing fine
- amber – we will need help soon
- red – help needed.

This is one way to help how children can be in charge of their learning while outdoors, and make decisions about when they need teacher support.

Like the Outdoors Science Boxes the contents of the haversack can be changed to include a focus on a particular aspect of science learning, for example, changing the haversack to an 'Outdoors ICT Haversack'. This could contain equipment for collecting and recording data and experiences, such as:

- a portable digital microscope
- Easispeak microphones
- digital camera
- video recorder.

What other resources can we put outdoors to support science?
A range of ideas and practice has emerged from the schools in the project. Each school developed approaches according to what they wanted to achieve. Some schools have used the idea of fixing boards on outside walls so that children and teachers can use these as a focal point for different activities.

Wall-mounted chalk and white boards

Grenoside Primary School decided that they would attach wall mountable white boards outside so that children could sketch on them, record observations of birds or invertebrates or even write their own science questions.

Opportunities for developing Personal Capabilities can be found in different contexts. Here the boards offer occasions when children can make their own decisions about what to record because they have ownership of the boards, not the teacher.

Teachers were creative in their approaches and were keen to support not just the children but also other teachers and adults. For this reason a number of teachers displayed activity ideas around the school grounds which served the dual purpose of engaging children and prompting and reminding teachers of the potential of different areas outside.

Outdoor activity displays

Sue at Cheveley Park Primary School created and laminated a set of activities and questions which were located at strategic points around the school grounds.

Sue used visual cues by taking photographs of children engaged in activities such as colour matching in nature, using pooters to collect invertebrates and binoculars to spot birds. This supported children, enabling them to 'read' the pictures or words so that they would be able to carry out the same activity themselves without the need for adult help or supervision. It also provided teachers with ideas for links into their science topics, and also supported them in sending children to carry out specific activities which were already developed and displayed.

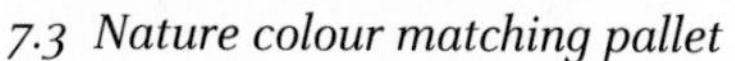
7.3 Nature colour matching pallet

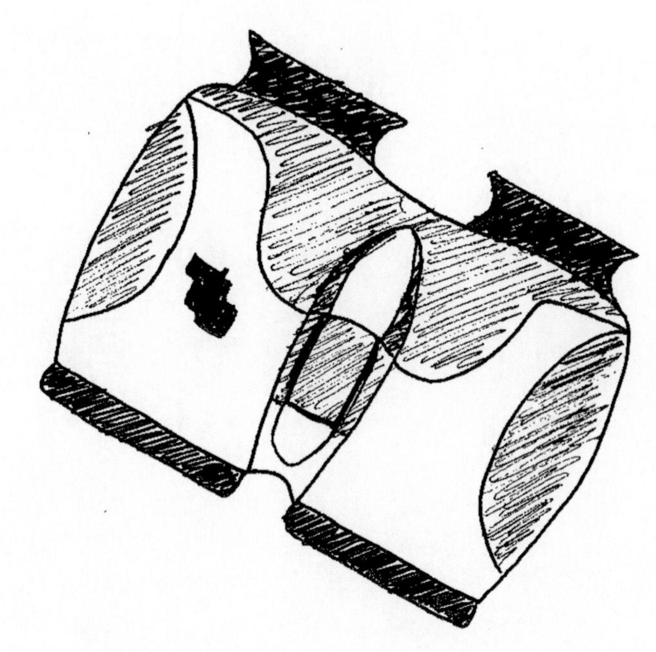
7.4 Binoculars

What will you do?

The most successful resources are those that are part of a well-thought-out plan, based on the resources already available and the key goals for developing science 'Beyond the Classroom Boundaries'. You might find it useful to think about the following:

- What are you trying to achieve by developing resources boxes or haversacks?
- What system of managing resources for science outdoors would best suit your children and staff?
- What do you already have available in school that could be used in the resource kits?
- Which resources would you, your staff and the children like purchased?

Before purchasing any resources think about:

- contacting your secondary school to borrow resources on a short- or long-term loan system
- asking local businesses for support (e.g. clipboards)
- using friends of school to fund-raise
- sending letters home to see what parents can donate (e.g. haversacks).

Practical ideas for teaching light 'Beyond the Classroom Boundaries'

Working 'Beyond the Classroom Boundaries' in science offers a range of opportunities to develop children's understanding of light and shadows as well as the prospect of helping to progress children's Personal Capabilities. Here are some ideas that you might like to try out.

Shadows around the school

Challenge children to choose six interesting shadows in the school grounds: large ones, small ones, strange shadows and shadows that are easily recognizable. Ask children to think about what they will need to take with them to record the shadows in the 15 minutes that they have to complete the task.

Children could decide to photograph the different shadows or draw them; this provides a great opportunity to create a shadow photographic or visual exhibition in the school grounds. Tell children that of the six shadows they can choose to exhibit pictures/photographs, which they can place in a plastic wallet, label with their name/s as the photographer and hang on a railing or fence. Ask the children to decide what kind of questions they could place alongside the photographs to encourage other children in the school to engage with the exhibition. This self-directed task would be ideal for groups of three, encouraging discussion and negotiation between the children and honing their decision-making skills, especially when challenged by the timed deadline.

Making shadows

Making shadows demands that children understand how shadows are made, especially when the children are challenged to:

- make shapes with their own shadow
- work in pairs to make action shadows
- join with another pair; one pair makes a shadow and the other pair works out what the shadow represents
- plan, organize and carry out a mini shadow drama using their own shadows.

Use strategies such as pair with pair, where children work together, and then share what they are doing with the other pair, who have to guess what the shadow is, what it is doing etc. This task would be ideal to link with cultural stories in literacy and religious education, enhancing a cross-curricular topic with creative expression and critical thinking.

Changing shadows

For this activity children are asked to answer the following question: 'How do different shadows in the school grounds change during the day?' It is sometimes useful to deliberately organize children into larger groups: for example, groups of six, particularly where activities require children to do a lot of work within a short space of time. The question states that they have to find out how different shadows change in the school ground during the day. When broken down this will require children to:

- choose different objects around the school that make a shadow
- decide how often they will need to go outside to monitor how the shadows change.

The teacher then challenges the group by indicating that every time they go out they have a time limit of 10 minutes. It is to be hoped that the children will realize that six individuals measuring one shadow at a time is not good time or people management. If children work in pairs, and each pair has one or more shadows to measure then the 10 minutes allocated should be fine. Further questions can be asked to help the children organize themselves to carry out the activity, for example:

- How will you know if the shadow has changed during the day?
- What do you think you will have to do to answer this question?
- How will you organize yourselves to get the task completed in the 10 minutes you have been given?
- What will you need in your haversack to help you carry out this activity?

Children might choose any of the following to put in their science haversacks:

- tape measure
- metre ruler
- stopwatch/clock
- chalk
- masking tape
- Easispeak microphone
- individual whiteboard, whiteboard pen and duster
- digital camera
- clip board, paper and pen.

The key from the perspective of developing Personal Capabilities is to scaffold the problem-solving process and assist the planning and organization of the task so that the larger groups achieve success. The use of Role Badges can be of help, as well as thinking frames: for example, GRASP (see Chapter 2, p. 22) assisting the teams in collaboratively thinking their way through a lengthier task.

Modelling shadows

A very good way to elicit children's understanding of how shadows are made is to ask children to model or create a physical enactment of how shadows are made. Tell the children that they have to make a 2-minute video clip to show other children in their class or in the school to help them understand shadows. Explain that they are allowed to create a commentary to go alongside the visuals in the video clip. Allowing children to work outdoors gives them the space to work unhindered by furniture or other groups engaged in the same activity and also it is more likely to encourage creative approaches. If different groups are challenged to adapt their explanations for different audiences, the essential skill of communicating messages in ways that can be understood by different groups of people can be discussed and reflected upon.

Map of light around the school grounds

In this activity children are challenged to create a map of the school grounds and show how light changes in six different places over a school day and to explain why the light levels change. If children are told that they are not allowed back into the classroom once they go outside and that they will have a set amount of time to complete the activity, then it helps to focus children's thinking on planning what they will need to take out with them. In this particular activity children usually have to think about the following in order to make sure they can work outside without returning to the classroom:

- how to use the haversacks to help organize themselves
- using a light meter or computer light sensor
- how to time-keep to make sure that they complete the task in the time given
- decide what else they might need to record their findings (e.g. camera, clipboard, paper, individual whiteboard, pens, duster).

This is another activity which could lend itself to a range of areas of development. In providing justified feedback to the group the children can be encouraged to support their thinking with evidence using a range of sources. Such skills are both scientific investigation skills as well as relevant Personal Capabilities.

How does the amount of light in different areas of the school grounds affect biodiversity?

One aspect of science that is very important is to develop children's ability to link their knowledge and understanding of what might appear to be disconnected areas of science. In this example, if children know how to measure light they could map light and plant growth to find out if there is any relationship between the amount of light and the type or number of plants. Again this is an activity that would benefit from children using Personal Capabilities related to self-management, such as organizing their approach, equipment and roles.

CHAPTER 8

Recording and communicating science 'Beyond the Classroom Boundaries'

What is the goal?

Project teachers frequently asked how can children record their learning outdoors; the common-sense approach to children recording and communicating 'Beyond the Classroom Boundaries' is to limit the use of paper, as soggy pieces of paper are no use to anyone. Realistically, we need to compile evidence of learning outdoors in ways that do not compromise the development of children's independence and the integrity of the science itself. We need to become more flexible and creative in our approach and broaden the children's repertoire in relation to recording and communicating when working 'Beyond the Classroom Boundaries'.

In primary science our aim is to make sure that by the end of the primary years children have had a wide range of opportunities to record and communicate their science, so that they have the confidence to communicate their science appropriately using scientific language, knowledge and understanding in a concise way:

- within a group
- to others in their class
- to different audiences beyond the classroom.

In this chapter we aim to share:

- creative approaches to recording learning when children are working 'Beyond the Classroom Boundaries'
- strategies for supporting children to become more independent in recording and communicating their science learning outdoors
- thoughts on how to encourage the Personal Capabilities of self-management and self-awareness while learning outdoors.

What is the reality?

The reality is that much of the recording and communication in primary science is paper-based and this is inappropriate when working outside. It is almost absurd to expect children to manage paper and pencils on a windy day outside with puddles on the ground. It is essential that we seek appropriate alternatives so that teachers can be confident there is evidence of learning from groups or individual children.

The expectations of most science curricula is that children should, by the end of the primary years, be able to make their own choices about the most appropriate ways to record and communicate their science. In relation to working 'Beyond the Classroom Boundaries', we need to gradually extend their ability, when working outside, to choose and use different ways to those that apply to the classroom environment.

In project schools the reality of recording learning while outdoors initially took second place to the experience itself. Teachers focused primarily on getting the children safely outdoors and providing motivating and structured activities for them to do. They independently started to audit their approaches to recording in science in the classroom, and considered what could be taken outside. They also sought to introduce new outdoor approaches, such as exploiting different forms of ICT, or using handheld whiteboards.

Initially, teachers started with digital cameras, since this seemed to be the most obvious way to record experiences outdoors and the one which most children were able to handle without teacher intervention. Gradually, teachers began to realize that not only did they need to introduce different ways of recording outdoors but they needed a discussion with the children about what the children thought was the best way to record their work outdoors. Teachers also began to encourage children to think about how best to communicate their outcomes, both to each other and when regrouping with the teacher and the rest of the class. Teachers from project schools began to show greater interest in what standard and technological resources they already had in their schools that could support recording and communicating science 'Outside the Classroom Boundaries'.

What are the options?

Recording learning 'Beyond the Classroom Boundaries'

There are many ways in which children can record their science when working outdoors, both formally and informally. Here are some ideas from project schools.

Science boxes

The science boxes described in Chapter 7 could contain pencils, crayons, chalks and watercolour paints, paper, individual white boards and clipboards for children to use mainly during play and lunchtimes. Children could also have access to plastic wallets and string so that they can put their work inside and display outdoors: for example, hanging the wallet from railings or low branches on bushes or a tree.

White boards

These can be used in different ways, such as fixing a large whiteboard to a wall in the school grounds or, for example, in a wildlife hide. Children use this to log observations such as bird sightings, and to place questions on or to log data. Alternatively, children could have handheld whiteboards to take out as a replacement for paper, allowing them to record observations and information using wipe-clean markers. The benefit of this method is that they are portable and water resistant, and information can be easily shared with the teacher and other groups working outside and when they are back inside the classroom.

Clipboards

When it is essential for children to record on paper, clipboards are a must; when children are given them to use they often feel as though they are real scientists. On a practical note it is useful to tie a pen or pencil onto the clipboard using string, and to attach a piece of overhead projector film to the board so that it can flip down to protect the paper from raindrops and mud!

Digital technology

Digital technology (ICT) for schools continues to advance and children are quick to learn to use new equipment for a range of purposes. Table 8.1 provides suggestions for how a range of equipment can be used by children as tools for recording their science when working outside and also for communicating to others.

Table 8.1 Suggestions for use of digital technologies 'Beyond the Classroom Boundaries'

ICT	Suggested uses
Computer data loggers	• *Light* – levels around the school, changes over day and longitudinal study – changes across months • *Sound* – levels around the school, changes over the day • *Temperature* – around school grounds, changes over the day and monthly patterns, comparisons and reasons, links between light level and temperature, temperature under trees, in pond • *Pulse meter* – children's pulse rates after different kinds of exercise, before, during and after playtimes • *Soil moisture sensor* – different areas of the school grounds (shaded, exposed, low-lying), across the seasons, links between plants and moisture content of ground (e.g. trees) • *Wind turbines* – to investigate real-world conditions in different parts of the school • *Solar panels* – to investigate real-world conditions in different parts of the school, link information to temperature data

ICT	Suggested uses
Digital cameras	Photographs – children working, trees across the seasons, wildlife, action pictures (e.g. children kicking ball in slow motion to show forces, type and direction)
Video cameras	Children working, role-playing, results of investigations, demonstrations, roving reporters
Easispeak microphones	Collecting sounds, recording observations e.g. investigations, where invertebrates are found, how birds on bird table move and feed, music from their own made instruments, roving reporter, creating their own animal sounds, record ideas to listen to back in the classroom
Web cameras	Could be used in the classroom or by the whole school in a central location or on the school website to watch birds in a nesting box
Digital Blue Head camera	A hands-free approach which enables children to video as they work and move around school grounds
Easi-Scope	A computer microscope that can be used outdoors with a laptop; it enables children to view and store microscope images of objects e.g. leaves, soil, bark, a feather
Vi Tiny Digital Mobile Microscope	A mobile digital microscope that can be used without a laptop outdoors, the user can download images back in the classroom

A number of project schools decided to create their own ICT kits to support children working 'Beyond the Classroom Boundaries'. An important issue for the teachers was to make sure that the ICT equipment was available for the children to choose and use when they thought it would be useful to their activity, as opposed to rigidly directing their use. Such levels of independence are developed over time, and initial suggestions are necessary, as well as clear instructions for use and thoughtful reflection on outcomes. The teachers considered a number of ways of organizing this:

- making sure that the equipment was kept in their classroom close to the door so that children choose items as they went outside
- placing the ICT equipment in a special outdoors box
- creating an 'ICT Haversack' similar to the resources haversack suggested in Chapter 7.

Communicating learning 'Beyond the Classroom Boundaries'

Once children have recorded their outdoor science it is necessary to consider how they will communicate their ideas, ways of working, new knowledge and results to a range of audiences. For children to communicate their science successfully they need to develop an understanding of their audience, which means that they will need to consider:

- who their audience is
- what their audience needs to know
- the best way to share their learning with their audience.

Children will need to make sense of their experience and their results and be exposed to different audiences and ways of communicating. Making links to literacy are beneficial, for example, creating sentences and poems, writing newspapers and producing plays, or creating PowerPoint presentations, and this will certainly require children to consider organizing what they want to communicate.

Project schools rose to the challenge of finding different ways for children to record and communicate working 'Beyond the Classroom Boundaries' and developed a range of creative and innovative approaches which are shared here.

Communicating through Outdoor Display Areas

Sue Harrison of Cheveley Park Primary thought that if the children are working outside then the products of their work should be displayed outdoors. In her school this led to the challenge of where the children's work be placed and how to display it so it was not ruined by the weather. Her solution was to laminate and use plastic wallets (sealed with Sellotape and hung upside down) so that the rain did not spoil work. Gradually the school grounds were transformed through the development of outdoor displays of children's drawings, stories, poems and fact files, which was a mix of material that children had created during playtimes and during lessons outdoors.

Personal Capability: Positive Self-Image – valuing oneself and one's achievements.

It is a well-recognized fact that children gain a heightened sense of achievement and self-esteem when their work is publically displayed and recognized. Sue's commitment to inclusive displays – where each child had opportunity to share their work – was vital to achieving this. Children talked about their work, showed it to friends in playtimes and parents/carers at home time – this alone enhancing the learning processes as well as building positive self-image.

Sarah at Castleside Primary adopted Sue's idea and found that there were a number of places where the children could display their own work, and these included a wooden fence, tree branches and a wire mesh fence on which they could hang work and which they also used to weave, using natural materials found in their school grounds.

Other ideas developed to share outdoor learning included children using sound recording equipment, such as Easispeak microphones, to produce podcasts which were then made accessible to children, parents and school governors on the school website. Children created PowerPoint presentations which were shared in school assemblies and which were also shown to school governors. Other approaches included the following ideas.

Floorbooks

Floorbooks can take various forms, but they are usually large books, for example, A2 size, bound simply with string that contain a range of children's work which might include Post-it notes, captions of verbal thoughts, photographs, labels, sketches and writing. More information about these is available online at the AstraZeneca Science Teaching Trust (www.azteachscience.co.uk) where you can view the unit developed by Kendra McMahon (Senior Lecturer in primary education at Bath Spa University College) and Rhianedd Baker (primary teacher).

Photographic exhibition

The increasing simplicity of digital cameras allows even novice users to create good photographs, and children can and do rapidly become proficient users. If you have a member of staff, a parent or governor who is a keen photographer then do arrange for them to give a 'master class' to children to help them to develop a skill that can be used in primary science.

When working outdoors children can take photographs for recording:

- how they are working: for example, how they set up an investigation into 'Which ball bounces highest?'
- before and after sequences, such as floating and sinking objects
- observations of animals, trees: forces in action
- changes, such as seasons
- awe and wonder moments: for example, a beautiful flower, or a spider in a web sprinkled with dew.

Schools involved in Personal Capability development have explored the notion of 'PC Detectives' where children use cameras to capture images of particular skills in action: for example, teamwork, communication, collaboration, problem solving and organizing. Children can be encouraged to track their learning process and to consider photographs that tell the story of 'how' they learnt and 'how' they have worked as well as recording science outcomes.

Our aim in allowing children to store and sometimes print their science photographs should be to help children understand that photography is just another way that they can record and communicate in primary science, just like creating a table or writing a sentence. You might like to consider showing some examples of scientific photos such as X-rays and time lapse photography sequences and video clips to illustrate how these are used by scientists. Photographs and video sequences are important because they allow children to capture what has happened and take it indoors with them to analyse, compare for similarities and differences, record changes in the environment over time, and of course help to remind them what they did and what happened. In some instances, photographic evidence in science can be just as important as numerical evidence.

You might also like to consider organizing (or children arrange) regular exhibitions of science photographs which can serve a number of purposes:

- to inform other people about the science that the children do
- to celebrate science
- to celebrate children's ability to use ICT effectively
- used on information points around the school grounds
- as part of 'What is this?' school competitions and quizzes.

Some of the project schools decided to have a science photographic competition where scientific photographs were taken in the school grounds by the children across the year and the winning 12 photographs used to produce a school calendar.

Talking books

Schools in the project found that Talk Time Postcards, Talk Time message buttons, Talk Time Mirrors and Talk Time Books are invaluable. These resources allow children to record the spoken word for a given time and allow playback of what they have said. The process challenges children to:

- think through their ideas before recording
- be concise
- work to a time limit.

Below are some examples of how these can be used.

Children placed items such as leaves, feathers, seeds in the plastic wallet and then recorded a question such as 'Which tree does this leaf belong to?' 'Which bird lost this feather?' 'How is this seed dispersed?'

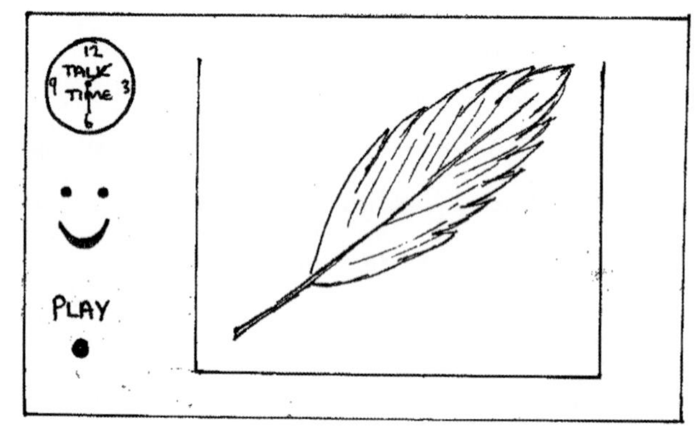

8.1 *Talk time postcard*

In each of the page wallets children place items relating to their work outside. They could include a data table, photographs and even a tree bark rubbing and record information, descriptions, explanations or even quiz questions relating to each of the pages.

8.2 *Talking photograph album*

Feasey (2005: 72) suggests that 'creative communication does not always require the use of the written word'. In the example below Kay at Wheatlands Primary used a wonderfully creative approach with her children to communicate information on the invertebrates they found in the school grounds.

8.3 *Clay tile invertebrate decision tree*

Clay tile invertebrate decision tree

'As part of exploring the animal life of the school grounds we decided to give the children the experience of recording their observations through making invertebrate clay tiles.

The tiles are about 10cm² and on each tile is the outline of an invertebrate found in the school grounds, such as an ant, beetle, woodlouse, bee, either in relief or indented into the clay. The children were challenged to make sure that their invertebrate outline was scientifically correct, that is, the outlines were in proportion and had correct body parts, for example, all of the insects had head, thorax and abdomen, whilst other invertebrates such as centipedes had one pair of legs per segment in contrast to the two pairs of legs per segment found on millipedes.

Once hardened the clay tiles were sealed using an outdoor varnish and then became part of an outdoor decision tree. Each tile was fixed low down on an outdoor wall, using 'No Nails Glue' so that the tree was accessible to all.

I was really pleased with the results, for a number of reasons: it provides children with a large hands on accessible decision tree outdoors, which is fun to use and one that can be added to as children explore and observe new invertebrates in the school grounds. The children can access it whenever they want, during lessons and play times; they can also help each other work out which invertebrate they have seen. Finally because the clay tiles had either indents or raised parts children were able to take 'crayon' rubbings of invertebrates that they had seen or collected and put their rubbings in a folder or book back in the classroom. Children are able to use paper and crayons from our Outdoors Science Box.'

Taking it further

'I was concerned about whether the decision tree would be used regularly, so I also developed activities for children to use the decision tree. I created problems for children to use the decision tree to find out, for example, "How many invertebrates are insects?"

'"What does a woodlouse look like? How many places in the school ground can you find woodlice?" These were laminated and placed next to the Clay Tile Decision Tree and changed on a regular basis. When children found an invertebrate and were able to identify it, they took crayons and paper from the Outdoor Box and did a 'Clay Tile' rubbing to place in their science book or file.'

Personal Capability – Communication: to communicate ideas, information and feelings in different ways.

This activity challenges children to share their observations in visual and kinaesthetic means. Kay talked to the children about ways to record and communicate the children's observations of the minibeast they had found. They took photographs initially of the bug, produced sketches and then the clay tile, and later wrote short descriptions and haiku poems about their bug. The children discussed the benefits of each form of communication – commenting that:

- "the photographs helped them see the colourings of the bugs, but that only through more detailed observation and sketches did each and every part of the bug become clear."
- "the clay tiles were fun to touch and rub, allowing them to play blindfold games to guess the bug."

Encouraging children to appreciate different forms of communication and the roles they play in science learning were the "smart" outcomes from this activity.

PE equipment

We are probably familiar with ideas such as using PE hoops so that children can use them to create sets and Venn diagrams when classifying objects: for example, when outdoors, children could use them to sort leaves found around the school grounds, deciding their own criteria for classification or using given ones such as compound or simple leaves.

In the example below children at Bradway Primary School were challenged to create a human skeleton using a wonderfully creative approach.

Outdoor Skeleton

Bradway School had usually done most of their work on the human body indoors. Since embarking on developing science 'Beyond the Classroom Boundaries' staff decided to try to take as many of the science activities outside as they could. When it came to the skeleton the teacher took a totally different approach. Normally, she would have provided the children with large sheets of paper and asked them to draw around themselves and put in the bones, or ask children to create a skeleton from recyclable materials.

This time the children were organized into small groups and given access to small PE equipment such as skipping ropes, bean bags, rounders bats, team bands and quoits.

When the lesson finished the children were really proud of their efforts and did not want to dismantle their skeletons. So they decided to leave them over playtime but felt the need to guard them from other children. What actually happened was that the children spent their playtime giving an impromptu master class on skeletons to the rest of the school!

Of course photographs were taken of all the skeletons to be placed in a class big book on the school website and on the screen in the school foyer.

Personal Capability – creativity: to be imaginative in sharing knowledge.

The first thought would have been to do this as a teamwork activity as they worked in groups of 6 and 7. However, the teacher was reluctant to give roles such as 'leader' as she did not want one child to dominate the group, or be seen to have a specific job. She wanted them all to be able to get hands on, talk with each other and be thoughtful and imaginative about what they did. She chose the groups in order to limit behaviour problems which resulted in all being on task and everyone enjoyed sharing learning in a large visual way.

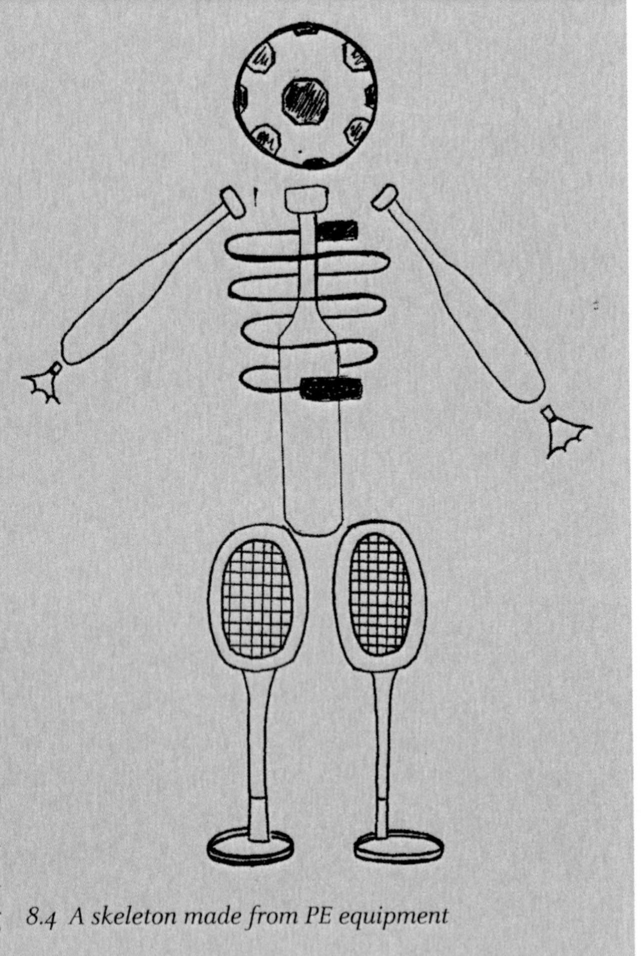

8.4 A skeleton made from PE equipment

This was a really innovative and motivating approach which the children thoroughly enjoyed, working outside and using the space to full effect. Sensibly the teacher had made sure that they had plenty of time to carry out this activity, which allowed them to think creatively, discuss their ideas and frequently change their skeleton to improve it until they were satisfied with their efforts. As they worked they discussed, shared their knowledge of the skeleton and argued about which items of PE equipment would be best for different parts of the body which often resulted in children changing their ideas and compromising.

What will you do?

A good starting point would be to take up the suggestion from earlier in this chapter and to audit the different methods that teachers use when their children record and communicate their science. An easy way to do this is to ask teachers to complete an online form, an example of which is in Table 8.2, which asks them to indicate how they would usually ask children to record and communicate their work and then to suggest an alternative suitable for outdoors.

Table 8.2 Auditing, recording and communicating science outdoors

Topic	Outdoors activities	How will children record?	How will children communicate?	Try this instead
Friction	Cars down a ramp	Record results using table in science book	Write up investigation	Record results using Easispeak microphone Take photographs of themselves working and how investigation is set up, print and annotate

Teachers could take their grid (last column left blank) to a staff inset session and then work collaboratively with colleagues to consider alternative approaches.

Gradually across the school year teachers will try out different approaches and therefore broaden their own and the children's repertoire in recording and communicating their outdoors science work. Do not forget to celebrate this work through the creation of big books, material placed on the school website or a plasma screen in the school reception area and of course do identify an area outside where children can display their own work.

Once you have audited the different approaches to recording and communicating science outdoors then the obvious next steps are to:

- identify issues
- support staff by offering additional approaches
- create a systematic year-by-year overview that shows how children progress in this area through the primary years.

Practical ideas for recording and communication 'Beyond the Classroom Boundaries' – invertebrates

Invertebrate Safari

In this activity children go on an 'Invertebrate Safari' during which they can be encouraged to get into role by wearing khaki shorts and a shirt and hat for the day. By telling the children that they will be out for several hours they need to organize and plan the task (self-management skills) and take all their equipment with them in their outdoors haversack. Give children time to think about what they will need to collect and to record what they find. For example:

- *For collecting invertebrates* – pooters or yoghurt pots and paint brushes for collecting, white container for observing, magnifying lenses, bug hut or container with holes for keeping invertebrates. Measuring equipment, quadrats, hoops, calculators to calculate population estimates.

- *For recording observations* – for example, what they look like, how they move and their habitat:
  - Easispeak microphone, digital or video camera.
  - light and temperature sensors to log data on conditions in the habitat
  - sketchbook and pencils to sketch their invertebrates using hand lens
  - invertebrate identification sheet or book.

Do not forget to tell the children that they must make a list of the equipment that they put in their group haversack so that they can check that they have everything when they return to class.

Communicating learning

When the children return to the classroom with all their information, they might need to carry out some research to 'fill in the gaps' such as life cycle, names of different parts of the body, and so on. Once they have all the information they need, there are many different ways that they could communicate their 'Invertebrate Safari', for example:

- Design and make *Invertebrate Information Points* that can be placed around the school grounds for others to use.
- Take the role of *Wildlife Rangers* who can take other children on an Invertebrate Safari.
- Design and produce a *PowerPoint Presentation* on their invertebrate to share with the rest of the class.
- Produce a *Two-minute TV Programme* on invertebrates in the school grounds for other classes to use.
- Produce a *Podcast* for the school website on their Invertebrate Safari.
- Contribute to an *Invertebrate Big Book.*
- Create a *Biodiversity in the School Grounds Leaflet* for children, parents and school governors, and send it to a local wildlife organization.
- Create your own *Clay Tile Invertebrate Decision Tree.*

Reference

Feasey, R. (2005) *Creative Science: Achieiving the WOW factor with 5–11 Year Olds.* London: David Fulton Publishers.

CHAPTER 9

Darwin's Thinking Path 'Beyond the Classroom Boundaries'

What is the goal?

When many of the schools involved with this project began exploring science 'Beyond the Classroom Boundaries', they were also celebrating the two-hundredth birthday of Charles Darwin. This sparked the imagination of both teachers and pupils and resulted in the development of a 'Darwin Thinking Path' in a number of schools. This path was based on Darwin's 'Thinking Path' which he created at Down House in Kent where he and his family lived, and where he wrote his most famous books *The Origin of Species* and *The Descent of Man*.

9.1 Down House

Darwin created a path which traced the outskirts of the grounds of Down House, passing by hedgerows, a field and through woodland. He called this his 'Thinking Path' which was also known as the 'Sand Path'. Darwin's life was dominated by routine: he walked the path every day for an hour from midday, enjoying the physical exercise and solitude which allowed him space to think. His use of the path tells us a lot about how Darwin as a scientist worked: each day he would take time out to think, undisturbed, as he walked round his Thinking Path.

> No matter what the weather might be like he went for a stroll at the Sand Path with his dog Polly, a white terrier. Along the way he would often stop by the greenhouse to check up on how his plant experiments were doing.
>
> (www.aboutdarwin.com/darwin/CD_Daily.html accessed 23 July 2010)

As he paced he mulled over his ideas, observations and results of his experiments, trying to make sense of them and making connections between different ideas and types of knowledge. The fresh air, gentle exercise and rhythm of walking helped his thinking processes and it was during these walks that Darwin began to make sense of his personal theories that became the basis of his writings.

Today Neuro-Linguistic Programming (NLP) recognizes that there is a link between mental processes and physical movement.

> Repetitive physical movements and activities involving major muscle groups (such as walking, swimming, biking, playing tennis, etc.) influence our overall state of mind, and thus provide a more general context for our thinking processes.
>
> (www.nlpu.com/Articles/article6.htm accessed 15 February 2010)

9.2 A Darwin Thinking Path

What is the reality?

In the schools that have developed a 'Darwin Thinking Path' teachers have recognized the benefits that having a designated Thinking Path can offer children when working in science, for example:

- linking physical exercise with cognitive processes
- giving children time out and space to think
- showing children that thinking space is important
- understanding that in order for children to be creative, they need 'time to engage in critical reflection'
- understanding that being outdoors can help to stimulate ideas and encourage children to make connections.

> 'It is better because it gets hot and stuffy inside and it is nice to get fresh air, it feels like play, but it is really learning and an extra break really, which is better because it is like a break during learning.'
>
> (Elliot, age 11)

An important aspect of the development of the Darwin Thinking Path in project schools has been how teachers have helped children to understand the link between their path and Darwin, so that they develop an understanding that they are modelling how a scientist worked.

What are the options?

So how can a Darwin Thinking Path be used to support children working 'Beyond the Classroom Boundaries'?

Children need time out – they require time to think about their science – Sternberg (1999: 25) quotes Bethune who, as long ago as 1837, suggested that a creative person can 'store away ideas for future combinations' but what is required is the opportunity and space for an individual to take time out of a busy classroom and curriculum to make those combinations or connections in their thinking. As Feasey (2005: 22) suggests a 'thinking environment is characterised by opportunities for the children to have time to engage in critical reflection, so that a gestation period might help children to form their ideas' just as Darwin needed time to think and be creative.

Giving children time out to think about their science, make connections, and talk with a fellow learner as they walk the Darwin Thinking Path might be thought as a luxury, but should be an integral part of science. It is therefore important that children have time to think through ideas and appreciate that, like Darwin, they can take inspiration from nature to help them understand their world. It is also important that children understand that like Darwin, they too can have access to fresh air, space and time out from the confines of the classroom to think and to have ideas.

Children might tread this path for a range of purposes – for example, to:

- think through their ideas
- plan investigations
- observe using all the senses
- note changes in the environment
- problem solve
- clarify ideas and understanding.

Setting up the path

One of the wonderful things about using the school grounds is that the resource is already there and, as we have said before, it is free. Setting up a Darwin Thinking Path does not require any expenditure, but it does need everyone, both staff and children, to take ownership and suggest ideas of how it will be identified and used. Below are ideas from different schools that have set up their own Darwin Thinking Path.

Castleside Primary School's Science Thinking Path

Sarah's Darwin Thinking Path was created around the perimeter of the school grounds but avoiding the Foundation / Key Stage 1 area, since it was thought that the older pupils would be distracted by children in the early years area.

The path has a starting point which is a special 'Darwin Stepping Stone' to denote the beginning of the Thinking Path. The stone was designed by pupils as a response to a school competition, the winning design placed onto the stone.

Around the Darwin Thinking Path staff and pupils decided to place words which would help the thinking process and they were created as temporary 'waymarkers' by the ICT club. The idea was that these would be changed regularly and new words displayed.

Suggestions of words from staff and children included:

Absorb	Believe	Cogitate	Concentrate
Create	Daydream	Deliberate	Doodle
Drift	Energize	Feel	Gaze
Hope	Hypothesize	Imagine	Journal
Listen	Love	Nurture	Observe
Pause	Ponder	Puzzle	Reflect
Serendipity	Share	Smell	Speculate
Think	Touch	Understand	View
Wander	Wonder		

As a result of wandering a Darwin Thinking Path, Sophie, aged 10, wrote:

Different leaves
There are different kinds of leaves like. . . .
bristled tipped,
doubled toothed,
triangular shaped,
heart shaped.

Spikey leaves,
finger leaves,
feather like,
broad leaves,
thin leaves, and
forever green they are.

Shaw Primary School's Thinking Path

'The idea originally came from 'The Great Plant Hunt' pack from Kew Gardens which was sent to schools. We already had a path around the perimeter of the school which was not used very much so we wanted to make this walk into an area that had some science zones that the children could interact with starting with a foundation stone and then the path which will include a perfume garden with plants with a very strong scent that attract insects, a mountain area with large pieces of limestone, sandstone, granite so that children can see how they weather and use them when we do work on rocks and soil, such as to see how they are scratched, easily worn and interact with them rather than just small samples on table inside the classroom. We have already got a kitchen garden so we do not want that on the Thinking Path, we hope to have an arid zone with plants suited to dry conditions and I have an idea that maybe we could have some of the beds raised in the shapes of insect such as a butterfly shape.

'We also intend to add areas on the Thinking Path that have a focus on other areas of the science curriculum such as sound, materials and forces because I want staff and children to explore all areas of science. We do have, in other areas of the school, such as the playground, equipment which is used for materials and forces activities but it would be good to extend the opportunities further.

'The head teacher and rest of staff are keen to have the Tibetan style flags as part of the Path. We want to screen print them and put different words on and each flag to show a different aspect. It would be brilliant visually and the sound the flags would make in different kinds of weather would be quite an experience. This would require some finance and we also want to invite an "Artist in Residence" to work with children and staff on this project. It all takes time, some things will cost, other things we hope for donations or to make things in school, when it is completed it will be a fabulous resource for science and indeed other areas of the curriculum.'

We must not forget though that taking time out to 'cogitate', 'muse', enjoy and be inspired by nature is also important to children's spiritual development and we should aim to offer children opportunities to experience awe and wonder of the world around them. Eight-year-old Alfie sums up this idea when he says:

You could learn nature and noises you can hear and what shapes and pictures clouds can make, like pigs or teapots.

9.3 Cloud-pigs and teapot cloud shapes

What will you do?

You could begin by sharing the idea of the Darwin Thinking Path with staff, children, school governors and parents: for example, you could show a clip from the film *Creation* where Darwin is walking his path. Some schools have held a special school assembly (outside) to launch the idea of the Darwin Thinking Path and invitations have been sent out to friends of the school explaining what will happen and also request help in setting it up. When ready to use, celebrate the launch of the path with a Thinking Path special event.

In the meantime in order to set up the path there are a number of questions that need to be considered, such as:

- Where in your school grounds could you have a Darwin Thinking Path?
- How can you ensure that staff understand the philosophy behind the Thinking Path?
- How will staff make sure that the children are given sufficient time for the path, so that they can observe, reflect, discuss, think, share ideas and collect information (e.g. data, photographs).
- When will you take staff round the Thinking Path so that they can decide how to use it with their own class?
- How will you and your staff explain to children the purpose of the Thinking Path?
- How would you encourage staff and children to take ownership of the path?
- What could be placed around the Thinking Path to encourage and support its use?
- How would you maintain the Thinking Path encouraging children and teachers to use it on a regular basis?

Practical ideas for using 'Darwin's Thinking Path'

Darwin walked his Thinking Path on a regular basis; while we do not expect children to walk the path for an hour every day, we would expect children to use the path on a regular basis. Use of the Thinking Path will vary but might include:

- children walking the Thinking Path in their own time during playtimes
- when they need some time out to reflect on a problem in a science lesson
- at regular intervals (e.g. once a week, fortnight or month) to note changes in the environment
- as part of a science lesson (e.g. habitats, sound, light).

When children use the Thinking Path they could be encouraged to observe and think in relation to any of the areas explored below.

What have you noticed?

- What have you noticed on your path today that you have not seen before?
- What is the light like today? How does it affect the Darwin Thinking Path? How does it make you feel?

- How many new plants can you see? How many invertebrates can you spot today?
- Talk into the Easispeak microphone and describe what you can see, hear, smell and touch while you are walking the path. Explain how you feel.
- Make a note of what is new on the Darwin Thinking Path board.
- Stop and take a photograph, store or print out and annotate with information; display it somewhere along the Darwin Thinking Path.
- Stop and take a photograph at the same spot every two weeks to record changes over a school year.

Similarities and differences

- How can you map and measure the similarities and differences in light and sound at different points along the Thinking Path?
- What are the similarities and differences between different parts of the Thinking Path?
- How many different 'micro-habitats' are there along the path, e.g. hedgerow, wall, under a tree, on tree trunk, under logs.
- Biodiversity along the Darwin Thinking Path: find out how many different plants and animals live along the path. Do different plants and animals live in different parts of the path?

Plants

- How many different plants are there along the Thinking Path?
- Create a class/school sketchbook of the plants found along the path. Make notes to go with each sketch, e.g. position, light, shade, damp, dry, dimensions (e.g. height, width), leaf shape and pattern.
- Identify and name plants.
- Design and make a plant label for each plant that will withstand the weather, with the plant's everyday and Latin name.
- Research stories behind the plants: for example, the Shepherd's Purse plant gets its name because the seeds pockets of the plant looks like an old-fashioned leather purse.
- Sort the plants into groups, for example, those with colourful flowers (to attract insects for pollination) and those without (such as grass that does not need bright flowers because it is wind pollinated).
- Create a school plant identification chart or key for the Thinking Path.

Trees

- Identify trees along the Darwin Thinking Path.
- Collect information about each tree, e.g. bark rubbing, leaf rubbings, twig sketches, tree silhouette sketch in different seasons.
- Collect numerical data about each tree, e.g. height of tree, trunk diameter, canopy width, distance of roots from trunk, average leaf size estimate of number of leaves on tree, light under tree.
- Work out the life cycle of each different tree: what do the flowers look like, what are the seeds like and how are they dispersed?

Seeds

- How many different kinds of seeds can you find? What kind of groups can you sort them into?
- How are the seeds dispersed?
- How far away from the 'parent' plant are the seeds found? How did the seeds get there?
- Set aside a piece of ground, leave it bare and see what grows? What kind of plants grow? Where did the seeds come from? How did they get there?
- Find out what a 'seed bank' is, and create a school seed bank, just like the one at Kew Gardens.

Animals

- How many different animals live along the Thinking Path?
- Observe and sketch the animals.
- Find out about each animal and its habitat.
- Create a food chain, a food web that contains the different animals.
- Why are the animals found there? Are the animals resident or visitors?

References

Feasey, R. (2005) *Creative Science: Achieving the WOW Factor with 5–11 Year Olds*. London: David Fulton Publishers.
Sternberg, R. J. (1999) *Handbook of Creativity*. Cambridge: Cambridge University Press.
www.bbc.co.uk/history/historic_figures/darwin_charles.shtml
www.greatplanthunt.org
www.nlpu.com/Articles/article6.htm

CHAPTER

10 Beatrix Potter – 'Beyond the Classroom Boundaries'

What is the goal?

In this chapter we consider how another scientist, Helen Beatrix Potter, can be used as a role model for both teachers and children when working 'Beyond the Classroom Boundaries' and explore how knowing about Beatrix Potter can help us introduce children in observing, collecting and recording plants and animals in the environment.

Beatrix Potter is most famous for her children's books with wonderful characters such as Mrs Tiggywinkle, Benjamin Bunny and Peter Rabbit. She was born on 28 July 1866 and died 22nd December 1943 and spent her early years holidaying in Scotland and the Lake District, where she developed her love of plants, animals and landscapes. Most importantly she was someone who carefully observed her surroundings and was able to translate her observations into detailed and accurate drawings and watercolours. Beatrix Potter also became a well-respected expert mycologist (someone who studies fungi); indeed, she created several hundred watercolours of fungi and wrote papers on germination of fungi. Later in life she became a farmer and avid conservationist, helping to form the landscape of the Lake District that we are familiar with today (Potter, 2006).

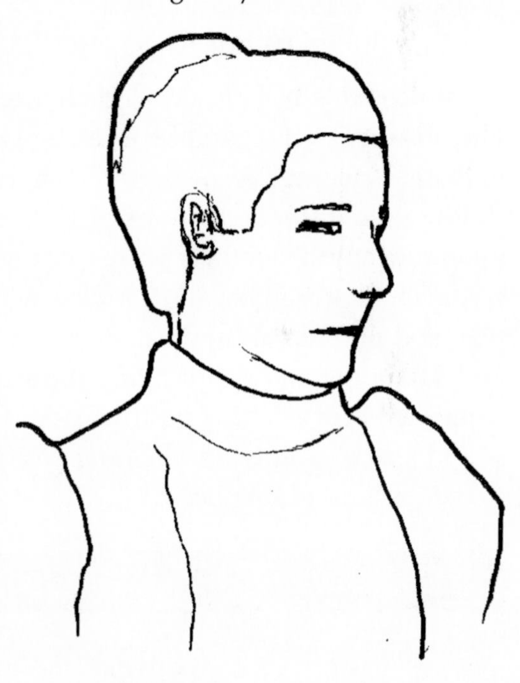

10.1 Beatrix Potter

What is the reality?

Beatrix Potter lived in a different era, one which allowed her to spend time observing, being curious, sketching and painting. Her vivid imagination led her to translate what she saw and sketched into stories which have captivated children and adults across the world. Today the pace in both everyday life and school can be fast and furious – this poem by William Henry Davies reminds us that we should take time to stop, observe, become familiar with and enjoy our surroundings:

> *What is this life if, full of care,*
> *We have no time to stand and stare.*
> *No time to stand beneath the boughs*
> *And stare as long as sheep or cows.*
> *No time to see, when woods we pass,*
> *Where squirrels hide their nuts in grass.*
> *No time to see, in broad daylight,*
> *Streams full of stars, like skies at night.*
> *No time to turn at Beauty's glance,*
> *And watch her feet, how they can dance.*
> *No time to wait till her mouth can*
> *Enrich that smile her eyes began.*
> *A poor life this if, full of care,*
> *We have no time to stand and stare.*
> (http://www.davidpbrown.co.uk/poetry/william-henry-davies.html accessed 23 July 2010)

What does this have to do with children working 'Beyond the Classroom Boundaries'? The answer is quite simple. In an age where many children spend much of their time indoors, watching TV or on the computer, school may be one of the few places where children are introduced to the process of taking time to stop, stand, observe and reflect upon the input from their senses. Primary science is the obvious place for this approach, where children can become familiar with the environment in which they live, work and play and develop an understanding of, and concern for, their natural environment. In understanding and appreciating the school's natural environment then children can be supported in developing positive attitudes related to stewardship, which, in turn, we hope would lead to children developing positive attitudes towards a sustainable future as they move towards adulthood.

Worms

Children at Woolley Special School were working outside one day when the children found worms. The interest in this lasted nearly a whole afternoon, which was unusual for this group of children. Rather than interrupt their interest, by either taking them back to the classroom at the end of the time allocated to science or redirecting their attention to the original outcomes of the lesson, the teachers allowed the children to lead the learning.

This group of children, who normally had a short attention span, were focused for several hours. While observing the worms, the children experienced a range of emotions from being fascinated to frightened. They were so excited that they were continually showing things to other children, which again was unusual for this group.

Their level of concentration was surprisingly high, and they were desperate to share their observations with teachers and other adults. Many of the children had manipulative as well as cognitive challenges, in the classroom they struggled with fine motor skills. Yet outdoors, given time to stop and stare at the worms, they were surprisingly dexterous and very gentle with them, handling them with real tenderness.

In primary science we need to make time for children to 'take time' to observe and get to know their surroundings so that children can:

- observe
- learn how to handle living things with respect and care
- make connections
- understand life cycles
- understand relationships in the environment
- know how humans can impact on the environment
- develop the emotions of awe and wonder.

An important element of science and working outside the classroom boundaries is to not be afraid to offer children time – and to do so with confidence knowing that the learning opportunities will be richer and deeper as a consequence.

What are the options?

In this section we share how several schools have considered how to use Beatrix Potter as the focus of different aspects of work in science linked with other areas of the curriculum such as design technology, literacy and art. The approaches taken by the teachers show how children can be offered innovative and creative learning opportunities 'Beyond the Classroom Boundaries'.

Sitooteries – Beatrix Potter observation points

If children are going to spend more of their time outdoors then they will need places to 'stand and stare' or to sit, so that they can think, observe and sketch. 'Sitooteries' is a lovely word meaning somewhere to sit outside, and many of the project schools have taken up the idea of having sitooteries in the school grounds. For some children it will be a luxury to be able to sit, relax and think, or to chat to a friend at play- and lunchtimes. Like so many seating areas around a school children enjoy the ambience that they can create. So how are sitooteries different to seating areas already in use in the school grounds. Well the sitooterie is special, because it is somewhere that children sit to think about science or to observe the environment around them. To make the science sitooterie different from just any seat outside we have to consider:

- making it look different
- creating a shared understanding of its purpose and use
- celebrating its use
- engaging the children in deciding where to place their 'sitooteries'
- engaging the children in the design and construction of a sitooterie (either a new one or changing an existing seat).

The location of the sitooteries is quite important. Children at Shaw Primary School helped to make decisions about where to place the sitooterie. They had to answer a number of questions by collecting evidence including:

- using sensors to find out light and noise levels around the school grounds
- finding places of interest (e.g. able to observe different types of plants, invertebrates, birds)
- finding interesting views, for example, between trees.

Children can use a range of digital technologies from computer sensors that enable them to collect and contrast data from around the school grounds to digital cameras to take photographs from a range of perspectives to help make decisions about interesting places to sit and observe.

When you and the children are considering where to site a sitooterie think about placing it near a point of interest, where the children will have things to observe (using touch, sight, smell and hearing): for example, a tree, flower bed or a wild area. If you are going to have more than one sitooterie make sure that some of the sitooteries are placed apart and others close together so that children can chose how to work, either independently or with a science partner. Some sitooteries might be located in a sunny spot while others should be in the shade and beside the seat could be a signpost which might include:

- information about animals and plants near the sitooterie
- suggestions for what to observe
- words such as 'cogitate', 'muse', 'ruminate', 'mull over' and 'ponder'
- activities such as 'How many different shades of green can you see?' and 'How many invertebrates can you spot and name?'
- activity prompts such as 'What will you sketch today?'

Once the sitooteries have been put into position why not celebrate by having a school assembly outside to introduce them to the children and gather ideas about how they might be used? Then ensure that each class builds opportunities for children to use the sitooteries into their schemes of work for science and individual lessons.

Sitooterie at Shaw Primary School

Carol and Haley, teachers at Shaw, liked the idea of a 'sitooterie'. Here they describe how their project developed.

'We were very keen to develop one in the school grounds. The children were challenged to design sitooteries for the school grounds, the children had to consider a number of issues:

- where to locate each sitooterie – quiet, noisy, sunny sheltered, an observation point
- design theme
- weatherproof.

'It began with the Year 5/6 children looking at existing photos of sitooteries and discussing why they made effective outdoor seating areas and where it should be positioned in the school grounds. The children became excited about creating one in our school grounds and already had lots of creative ideas.

The children then visited the proposed site for the sitooterie and began to formulate ideas as to what shape it might take. They worked in groups of three in order to piece together a design, working sometimes in the classroom and sometimes outdoors at the site. They really enjoyed having the freedom to visit the site when they thought it necessary to measure and check which plants would grow in the conditions and which sized furniture would be feasible.

The initial designs were in a rough format. The children then joined together with another group of three in order to tell each other about their designs and provide constructive feedback for each other. This worked very well indeed. The children were engaged in listening to design ideas and then suggested pertinent ideas relating to areas of the sitooterie which they thought required rethinking or improvement.

Following this feedback, the children in each group made any amendments they felt were necessary and then made a bird's eye view plan of their final design. In addition to this, the children also drew pictures looking at the sitooterie from different angles and showed the different features it had to offer.

We were delighted by how well-thought-out and creative the children's designs were! They had considered the purpose of a sitooterie and the needs and safety of all age ranges in the school.

Once the sitooterie design was completed, the children's next task was to prepare a presentation using their design and pictures to showcase in front of the rest of the class. The purpose of the presentation was to persuade the rest of the group that theirs was the best design! The following day, the presentations took place and the rest of the groups were given the very important role of awarding marks for certain aspects of the design. There was a lot to take in from the presentations so the children were also given the opportunity to wander around and look at the designs afterwards as well. The marks were then totalled. Interestingly, the three most popular designs which were all awarded the same mark!

The children were very excited to discover the winning designs. They seemed pleased that aspects of three different designs would be used in the sitooterie which has already been built in the shape of an S for Shaw. Now we are at the stage of thinking about landscaping the area around the sitooterie ready for use by the children.'

Beatrix Potter Haversacks

Of course, if children are to understand that some scientists, such as botanists and entomologist, sit, observe and then sketch like Beatrix Potter, we need to support them in this venture. Children really love having their own personal sketchbook, which they can use whenever they want to and share their sketches on their own terms. Along with this in art or science lessons they should be taught how to sketch, suggestions for this can be found in the 'Practical ideas' section on pp. 124–8.

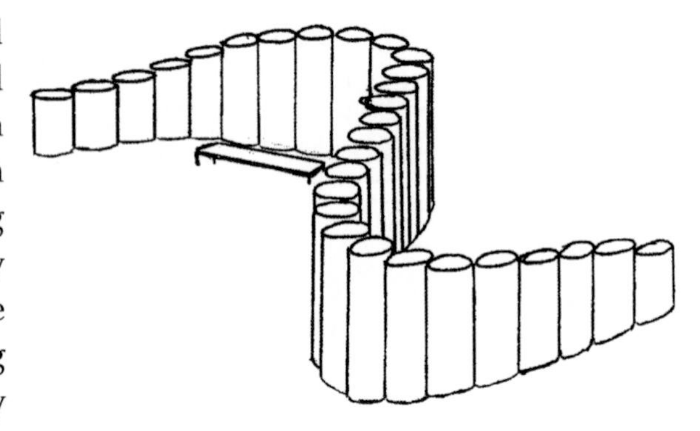

10.2 Shaw sitooterie

Of course, encouraging children to work like Beatrix Potter by sitting, observing and sketching is easier if children have the equipment to do so, and we can extend the idea of the 'Science Haversacks' from Chapter 7 by creating 'Beatrix Potter Haversacks'. Depending on the age and ability of the children the haversacks could contain any of the following: sketchbook, pencils, pencil crayons, pastels, small water colours palette, paintbrush and lidded container of water. You might even like to think about putting an easel near the sitooterie so that children can sketch or paint just like some artists do outdoors.

Displaying artwork

To encourage children to sketch in science do celebrate their efforts by allowing children to display their work outdoors on 'washing lines' or fences. They will soon get used to putting their work in plastic wallets and displaying them. In project schools children really enjoyed displaying their science artwork and quite often worked during a play- or lunch-time period and put their drawing in a plastic wallet and hung it in the outdoors 'science art gallery'. Of course, there is no reason why sketches and paintings done in either science sessions or art lessons should not be displayed in the outdoor gallery as well.

Alongside this your school could have special rewards for science artists and have science art exhibitions, with invitations to parents and friends of the school to view the exhibition. It really is important that part of the school policy for working in science 'Beyond the Classroom Boundaries' focuses on celebrating children's work in the outdoors, and staff should be encouraged to think about building this into their science planning.

Personal Capabilities

Working 'Beyond the Classroom Boundaries' is rich with opportunities that occur naturally and enable teachers to capitalize on developing children's Personal Capabilities. When children are engaged in using the sitooteries and sketching, there are a range of Personal Capabilities to be considered, as the examples below illustrate:

- *Creativity*: thinking of, sharing and playing with new or unusual ideas. In science, using art to explore the environment and using new techniques to record observations.
- *Communication*: communicating opinions and feeling appropriately. Children can use art and photography to explore how to communicate their observations.
- *Self-management*: when working from the sitooterie children are taking charge of their own learning, deciding when they want to sit, observe, sketch or paint. Being able to realize that they might only have a certain time to sketch and be prepared to put their sketch to one side and return to it at a later date requires *self-motivation*, being motivated to have a go and it also requires *tenacity, persistence and determination* to return to, for example, their sketch to complete it at another time and not just leave it half-done.
- *Positive self-image*: valuing oneself and one's achievements. Sketching offers an opportunity for children to share with other people by displaying their own work. This also requires the teacher to consider how to develop children's understanding of the ideas of self-image and their own part in developing the self-image of friends in the class by valuing the artwork that they display outdoors.

What will you do?

Of course, the reality is that our goal can only be achieved if first we value opportunities for children to stop and engage with the natural environment and, second, if we recognize that we need to offer children quality time. Sketching outdoors requires time to sit, observe, think and record what they see, something for which, in today's curriculum and indeed the world, time is limited. Giving children time is important, not just for what it can offer science but also the opportunities it provides for spiritual development, where children reflect on the beauty, intricacies and awe of the school's outdoor environment. Children will also need to be taught how to sketch and annotate to create scientific records of plants and animals (suggestions of how to do this are on pp. 126–8).

You might like to begin by thinking about using Beatrix Potter as a role model for the children: someone who enjoyed the outdoors, was curious and carefully observed animals and plants, then used those observations to create detailed sketches, stories and also to write scientific articles.

- How will you introduce children to Beatrix Potter, the scientist, artist and author?
- How will you develop children's understanding of the skills that Beatrix Potter had as a scientist and how she used them as an artist and an author?
- How will you encourage other members of staff to share Beatrix Potter and her work with the children?
- How will you encourage children to help design, make and position sitooteries around the school grounds?
- Where will children display their sketches?
- When will you have a scientific art exhibition?

Practical ideas for developing children's ability to sketch in science 'Beyond the Classroom Boundaries'

Observational drawing and sketching are important science skills, which we need to take time to develop so that children can be encouraged to record their observations while working outdoors. They are activities which support science in many different ways, for example:

- making systematic observations
- classification
- making comparisons
- using equipment such as magnifiers to enhance observations
- identifing simple patterns
- using scientific knowledge to make sense of their observations
- taking measurements.

Teaching children field sketching

1. The first thing children need to do is familiarize themselves with the overall shape of the object or animal. At this stage the children should not focus on detail – that will come at a later stage. One way of doing this is to ask the children to look at the outline of what they are going to sketch and then draw the outline in the air with their finger, all of the time keeping their eye on the object not their finger.
2. Ask the children to do this a number of times so that they become confident in creating the outline.
3. Then ask the children to draw the outline again but this time on the back of the person sitting next to them, with children taking it in turns. The friend can either guess what the object is or follow the outline saying, for example, 'straight line', 'curve' or ' wiggle'.

4 The partner describes the outline that they can feel and talks through the movements. 'Now you are going along the side, it slants a little, now the finger is going along a wiggle. . . .'

5 When children then take pencil to paper, we want them to do so with confidence and without using a pencil eraser too often. The aim is to scaffold the sketch by asking children to use the whole page and draw the outline of their object. Tell the children that they should keep their eye on the object and let their pencil do the work.

6 Once children have overall outline, then ask them to look at the next big feature and put that in: for example, when drawing a leaf the next big feature might be the stem running through the leaf.

10.3 Child drawing on back of other

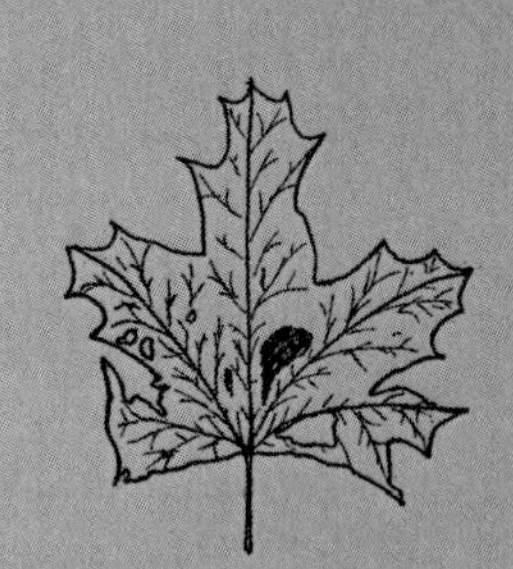

10.4 Sketching sequence of leaf detail

7 Each time the children have completed a feature they then focus on the next key feature, gradually adding more and more detail.

8 Once completed the choice is whether to add colour or leave it as a pencil sketch. If adding colour, pencil crayons are best as they allow for shading and fine detail.

9 Finally comes the labelling, with a title and then labelling key parts, and for this children should follow scientific convention which demands a ruler is used and that text should be written horizontally, at right angles to the sketch.

What can we do with sketches?

- add labels
- research information and add captions
- place it in plastic wallet and display outdoors

- place it in a class book on, for example, 'Invertebrates in our School Grounds', 'Our School Trees', 'The School Grounds Through The Seasons', 'Hidden Life in our School Grounds'
- create an indoor or outdoor exhibition
- laminate and place it as an information point at the appropriate place in the school grounds.

Using sketches to create stories in the style of Beatrix Potter

Beatrix Potter was able to bring her animal characters to life in her storybooks because she spent time observing animals and drawing detailed sketches and making notes. She observed how animals moved, their habits and habitats, as well as their quirky characteristics. Where children have studied animals in the school grounds they too could turn an invertebrate, bird or small mammal into a character and create a short story with illustrations in the style of Beatrix Potter.

The following suggests how to scaffold children to create Beatrix Potter type stories and illustrations.

1 Children begin by observing their invertebrate and researching the different parts and their names (e.g. segments, antennae, jointed legs, eyes, mouth, jaws).
2 They then draw a large version of their invertebrate: for example, a worm, with scientifically correct detail. Tell the children that they should draw their invertebrate so that it fills the page and not to draw it the same size as the real thing, because this would be too small to see the detail very well.
3 When they have completed their sketch tell the children that they are going to use their drawing to create a character for their own story. The first thing that they need to think about is what kind of character their invertebrate is: for example, happy, sad, bubbly, adventurous, curious, cheeky, posh or maybe serious.

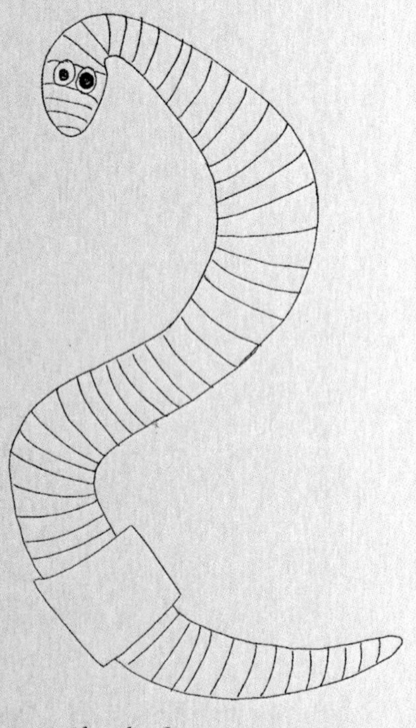

10.5 Sketch of worm

4 When they have decided, ask them to choose a hat, one that will help to show what kind of character the invertebrate has. Show children pictures of hats or a list, such as:

bowler hat	baseball cap	bonnet	top hat	miner's helmet
police helmet	flat cap	French beret	woolly hat	

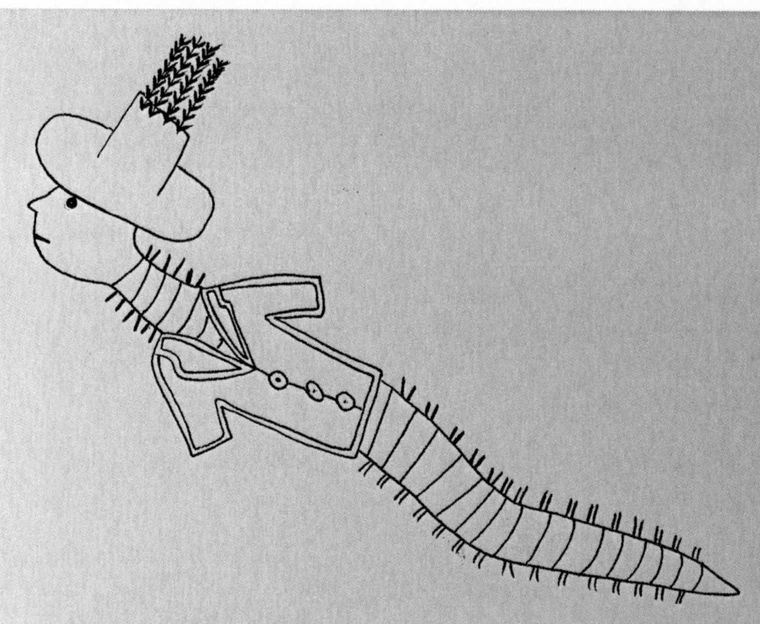

10.6 Millipede character

5 This helps to give the invertebrate a character; the next step is to draw the face.
6 After that the invertebrate can be given one or more items of clothing, such as a coat and gloves, that go with the hat and match the emerging character.
7 Finally, ask the children whether their invertebrate needs to carry anything such as a cane, miner's lamp, briefcase, haversack or basket.
8 Once the character has been drawn they can begin thinking about a mini story that is perhaps four–six sentences long. Their story should include the following:
   - introducing the character and setting the scene
   - explaining what has happened
   - explaining who else was involved (this could result in another character being created)
   - giving the character's response: for example, what did the worm do, say, and how did he do it or speak?
   - saying how the story ended.

Once completed there are a number of options, such as:

- creating a class book 'Class 5's Book of Beatrix Potter style Stories'
- reading their stories and showing their illustrations to other children using the sitooteries as story-telling seats or chairs
- creating a Beatrix Potter celebratory exhibition with information about the author and stories written by the children in the style of Beatrix Potter.

Illustration 10.7 is an example from Shaw Primary School of how children used sketches of animals found in the school grounds as the basis for short stories written in the style of Beatrix Potter.

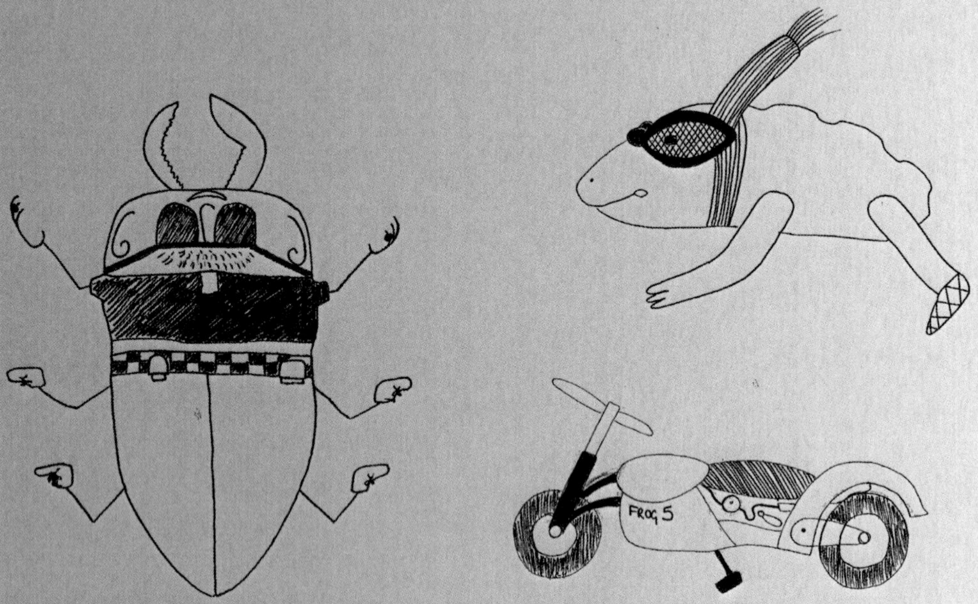

10.7 Story characters

References

Potter, B. (2006) *Beatrix Potter: A Journal.* London: Penguin.

CHAPTER

11 The future

The best way to predict the future is to build it.

(Douglas Adams; available online at www.futurelab.org.uk/resources/publications-reports-articles/opening-education-reports/Opening-Education-Report663)

It would be inappropriate for us to suggest what the future is for your school in developing science 'Beyond the Classroom Boundaries'. What we do hope is that science in your school is not constrained by physical and mental boundaries, and that teachers are able to work towards an open door policy, where science 'Beyond the Classroom Boundaries' becomes second nature to both staff and children. So, rather than try to fortell your future we have asked three teachers to suggest what they hope science 'Beyond the Classroom Boundaries' will look like in their schools. Here is what they have to say: we hope that along with the contents of the other chapters they will help to inspire and give you confidence that your future journey will be exciting and successful.

Kay Coverdale, Science Leader and Advanced Skills Teacher, Wheatlands Primary School, Redcar

'We have already begun the process of encouraging staff to take science outdoors into the school grounds. I have been working with a colleague this year and the rest of the staff have seen our children outdoors engaged in science activities. Gradually teachers have been looking at what we have been doing and tried things out themselves and we have watched as more and more children spend time outside doing science.

In a couple of years' time my aim is that we will be using the outdoor space for science regularly and no longer be "weather shy" . . . myself included! In fact I am going down to teach in Foundation Stage to toughen me up!

In terms of a longer term future I really aspire to having the ethos of Personal Capability development embedded in outdoors learning. It will mean going beyond what sometimes is a bit of show when children know the teacher is close by or watching, challenging them to use the capabilities and skills in proactive and reactive ways – not just thinking 'if the teacher sees me doing this she will be pleased'. The fact that being outdoors will be routine will mean that there isn't always someone right next to them watching and that they will be really 'doing it for themselves'. That would be a great outcome for our future practice, where the children use their Personal Capabilities because they see them as a life skill and useful, they have developed a strong sense of what they are, how to apply them and appreciate them as useful for getting the job done.

If we can achieve that then in three to five years I expect that staff will be planning bigger things outside, where the teachers have scaled things up both in size and time because they understand the freedom that the outdoors space can offer for science. I also think that it will be the norm to have children taking on outdoors science projects out of lesson time where they are involved in the planning process, questionnaires, getting research done and taking it to the head, who is open to suggestions from the children and willing for them to run things themselves.

I hope to have set the example for future practice in the construction of the clay tile "Decision Tree". I think of this as my legacy, something to be looked at, used and explored again over time to remind us of where we started and what our vision for the future could be.'

Carol Sampey, Deputy Head Teacher, Shaw Primary School, Wiltshire

'We have already moved quite a long way in the short time we have been engaged with this project, staff are using the outdoors more for science and children are beginning to make decisions about taking their science activities outside. The "SOLO Science Boxes" are well used and children have taken to these with great enthusiasm and the children have started to take ownership by being proactive – and suggesting what they want in the boxes Beyond the Classroom Boundaries'. Interestingly our learning science outdoor policy has already become 'outdoor learning policy' the staff decided that it should be a generic policy for across all curriculum areas which is a huge step forward.

Now I want to go back to original ideas and plans and really engage each year group so that they take on a project for the future, such as designing and developing different areas of the Thinking Path Science Zones.

On a daily level I think that in the future the children will be more aware of science, that it is all around them, there in the school grounds. I read somewhere, I'm not sure where, that science has had the greatest impact on the world and all its achievements are around us. I want children to appreciate this so that they are more aware of what things are like by really looking, rather than looking at things from books. I want children to be much more aware of what there is in their own environment not just in lesson times but things that are not part of the curriculum. I will know when this has been achieved when it is the norm for children to stop a friend or a teacher and drag them over to see something, or to try something out, for example to say how exquisite something is, like a sound, a spider's web or notice that there is a wasp flying around in November when it shouldn't be.

In five years time, I will be ready to retire and I would like to think that I have helped to leave a legacy in the Science Zones around our Darwin Thinking Path, which by then should be matured and well used. I hope that I will be able to watch children use the grounds throughout the day with limited but thoughtful support from staff. Ideally children will choose to do their science outside, use the sitooterie, think on the Thinking Path, sometimes during their play times as part of their relaxation.'

Grenoside Primary School

Our final contributions come from Grenoside Primary, one of the project schools and takes the form of two poems. The first is a collaborative poem written by Year 3 and 4 children following a science treasure hunt in the school grounds in September 2010.

Treasure Hunt
We're going on a treasure hunt, a treasure hunt, a treasure hunt,
We're going on a treasure hunt to see what we can find.
In my treasure chest I have placed:
A giant, juicy blackberry and a shimmering, golden flower,
A fly trapped in a cobweb for a spider to devour.
The Autumn sun upon my face and a glistening blackbird feather,
The giant clouds high in the sky that herald rainy weather.
We're going on a treasure hunt, a treasure hunt, a treasure hunt,
We're going on a treasure hunt to see what we can find.
In my treasure chest I have placed:
Some ripe, ruby-red berries and a leaf with big black spots,
Sparkling dew drops on the grass and a springy patch of moss,
A smooth, stripey pebble that's been polished to a gloss.

We're going on a treasure hunt, a treasure hunt, a treasure hunt,
We're going on a treasure hunt to see what we can find.
In my treasure chest I have placed:
Some jewel-like elderberries and the smell of wet tree bark,
A secret passage through the woods, mysterious and dark.
A delicate pink poppy and the gentle morning breeze,
Fluffy, ticklish dandelion seeds that made poor Harry sneeze!

We're going on a treasure hunt, a treasure hunt, a treasure hunt,
We're going on a treasure hunt to see what we can find.
In my treasure chest I have placed:
A mushroom like giraffe's skin and an oak leaf floating by,
A shiny, black berry like the pupil in my eye.
A crinkly, golden leaf and an all embracing calm,
A tough, armoured woodlouse that tickles in my palm.
We're going on a treasure hunt, a treasure hunt, a treasure hunt,
We're going on a treasure hunt to see what we can find.

The last word in this book is from Colin Fleetwood, the Head teacher at Grenoside Primary School, who has captured the essence of Science 'Beyond the Classroom Boundaries' in a poem:

Why Can't We Go Outside Miss?

Why can't we go outside Miss?
It's only just started to rain;
We've all got wellies and raincoats,
And fresh air is good for the brain.

Why can't we go outside Miss?
The air smells so good when it's damp;
We could take out some old scraps of plastic
And then build a nice cosy camp.

Why can't we go outside Miss?
It's wet so we're bound to see slugs;
The world doesn't halt 'cos it's raining,
We'd be sure to find lots of cool bugs.

Why can't we go outside Miss?
The raindrops are starting to dance;
We could measure how much rain is falling,
If only you'd give us a chance.

Why can't we go outside Miss?
We promise to do up our coats;
The puddles are getting enormous,
They're just right for sailing some boats.

Why can't we go outside Miss?
The clouds are all brooding and black;
The drainpipes are gushing like geysers,
If there's lightning we'll all come straight back.

Why can't we go outside Miss?
We can check out whose coat is the best;
And, just like you told us in science,
We'll make sure we do a fair test.

Why can't we go outside Miss?
We've all got waterproof skin;
We're happy to go on our own, Miss,
If you really have to stay in.

(C. Fleetwood, September 2010)

Reference

Bianchi, L. (2002) Teachers' experiences of the teaching of Personal Capabilities through the science curriculum, Ph.D. thesis, Sheffield Hallam University.

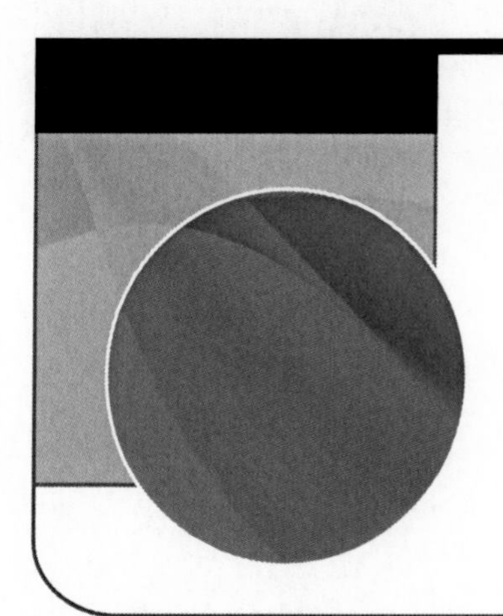

Index